THE ARYANS

A STUDY OF INDO-EUROPEAN ORIGINS

Owing to the antiquity and rarity of this book, certain pages are regrettably absent, and readability is compromised due to the blurring of the original text. If you possess any insights regarding the availability of these missing sections, kindly inform us. Your assistance is crucial in ensuring the completion of this significant work for the benefit of future generations.

THE ARYANS

A STUDY OF INDO-EUROPEAN ORIGINS

V. GORDON CHILDE

GYAN

Published by

GYAN

Gyan Publishing House
5, Ansari Road
Daryaganj, New Delhi-110002
Phone: 011-47034999, 9811692060
E-mail: books@gyanbooks.com

Distribution Network
gyanbooks.com
India, USA, Canada, UK, Australia

ISBN: 978-93-6780-510-7 (PB)

First Published, 1926

2nd Impression 2024

Printed at: Gyan Press, Delhi.

THE ARYANS A STUDY OF INDO-EUROPEAN ORIGINS
Author: V. GORDON CHILDE

The History of Civilization

Edited by C. K. Ogden, M.A.

The Aryans

The History of Civilization

In the Section of this Series devoted to PRE-HISTORY AND ANTIQUITY *are included the following volumes :—*

I. Introduction and Pre-History

*SOCIAL ORGANIZATION	W. H. R. Rivers
THE EARTH BEFORE HISTORY	E. Perrier
PREHISTORIC MAN	J. de Morgan
LANGUAGE: A LINGUISTIC INTRODUCTION TO HISTORY .	J. Vendryes
A GEOGRAPHICAL INTRODUCTION TO HISTORY . . .	L. Febvre
RACE AND HISTORY	E. Pittard
FROM TRIBE TO EMPIRE	A. Moret
*WOMAN'S PLACE IN SIMPLE SOCIETIES	J. L. Myres
*CYCLES IN HISTORY	J. L. Myres
*THE DIFFUSION OF CULTURE	G. Elliot Smith
*THE MIGRATION OF SYMBOLS	D. A. Mackenzie
*THE DAWN OF EUROPEAN CIVILIZATION	V. G. Childe
*THE ARYANS: A STUDY OF INDO-EUROPEAN ORIGINS .	V. G. Childe

II. The Early Empires

THE NILE AND EGYPTIAN CIVILIZATION	A. Moret
*COLOUR SYMBOLISM OF ANCIENT EGYPT . . .	D. A. Mackenzie
MESOPOTAMIA	L. Delaporte
THE ÆGEAN CIVILIZATION	G. Glotz

* An asterisk indicates that the volume does not form part of the French collection "L'Évolution de l'Humanité".

A full list of the SERIES *will be found at the end of this volume.*

The Aryans

A Study of Indo-European Origins

By

V. GORDON CHILDE

B.Litt. (Oxon),

Author of "The Dawn of European Civilization"

LONDON

KEGAN PAUL, TRENÇH, TRUBNER & CO., LTD.

NEW YORK: ALFRED A. KNOPF, INC.

1926

CONTENTS

LIST OF ILLUSTRATIONS

LIST OF PLATES

PREFACE

THE startling discoveries in the Ancient East and the great progress made in the study of the prehistoric civilizations of Europe, and especially of Greece, seem to make the moment propitious for a fresh survey of the fascinating question as to the origin and diffusion of those languages to which we, in common with the ancient Greeks, Romans, and Hindus, are heirs. In fact, no full discussion of the Aryan question has appeared in English for the last twenty-five years, while during that time the Minoan origins of the pre-Hellenic civilization of Greece, the presence of Aryan rulers in Mesopotamia by the XVth century and of an Indo-European element in the Hittite language have been revealed.

Yet my path is beset with pitfalls. Philologists will at once complain that the term "Aryan" is unscientific. Of course, I know that only the Indians and Iranians actually designated themselves by this name. But what expression is to be used conventionally to denote the linguistic ancestors of the Celts, Teutons, Romans, Hellenes, and Hindus if Aryan is to be restricted to the Indo-Iranians? The word Indo-European is clumsy and cannot even claim to be scientific now that Indian Sanskrit is no longer the most easterly member of the linguistic family known. Dr. Giles' term, Wiros, is certainly accurate, but, as thus written, it is so ugly that the reviewers have laughed it out of literature. Aryan on the other hand has the advantage of brevity and familiarity. I therefore propose to retain it, quite conventionally, in the traditional sense.

In the second place views on several crucial issues are very much in a state of flux at the moment. They may at any time be revolutionized by the fresh discoveries that are being announced every day from India and Cappadocia. Yet it is just this activity which makes an attempt to clarify the whole question urgently needed. To await the decipherment of all the Hittite archives and the excavation of every mound in the Indus valley would be cowardice. Still, the uncertainty ruling in these domains makes a relatively full statement of evidence desirable. That has been attempted in Chapters II and III. Without going into technical details, I have tried to summarize the main possibilities and to refer the reader to the chief sources of fuller information.

But the literary evidence from the Ancient East and the Ægean is still inconclusive. It must be supplemented by archæological and anthropological data. Hence in the subsequent chapters, the several traditional theories on the "cradle of the Aryans" have been re-examined in the light of the new evidence of that kind. But this procedure is peculiarly precarious. "Race" has different connotations for the physical anthropologist and the philologist. At the same time the correlation between the cultural groups, defined by pottery, tools, and weapons, and ethnic or linguistic groups is always speculative. It is only exceptionally that we find in a given area one culture superseded bodily by another in such a way that only ethnic movements will explain the change, and it is still rarer that the new element can be traced unambiguously to a specific focus. Normally other factors, such as trade and cultural borrowing or mere convergent evolution, have to be taken into account. Conversely a new racial or linguistic element may insinuate itself into a given province without producing any abrupt change in culture. As a science based upon abstraction and comparison, prehistoric archæology cannot aspire to the concreteness of history. Hence, while making every possible allowance for such disturbing factors, I have deliberately simplified—perhaps over-simplified—my account of the racial history of Europe and Asia rather than cumber these pages with a mass of technicalities which would still fall short of the complexity of the real.

It has seemed kindest to pass over in silence two theories recently propounded in England and France respectively, since they are so ill-founded that they will not even possess an interest as historical curiosities.

For the photographs illustrating this book, I am indebted to the courtesy of the Deutsch Orient-Gesellschaft (Plate I), the Trustees of the British Museum (Plates II and IV), Sir Flinders Petrie (Plates V and VII), the Director-General of Archæology in India (Plate VI), the Urgeschichtliches Forschungsinstitut, Tübingen (Plate VIII, 1), and the Royal Anthropological Institute of Great Britain and Ireland (Plate VIII, 2). I must also express my sincerest thanks to Miss M. Joachim for reading the proofs.

V. Gordon Childe.

TRANSLITERATIONS

The palatals are represented in Indo-European by *ḱ*, *ǵ*, *ǵh*. The corresponding sounds in Sanskrit are transcribed, according to the orthography of the *J.R.A.S.*, by *c*, *j*, *jh* and the palatal *s* is written *ś* (pronounced rather like *sh*).

The Sanskrit linguals are written *ṭ*, *ṭh*, *ḍ*, *ḍh*, *ṣ* (pronounced *sh*) and *ṇ*.

The Sanskrit anusvara, *ḥ*, is derived from final *s* or *r* which is, however, sometimes retained for clearness.

In Old Persian *š* is pronounced *sh* and so in other languages using the cuneiform script and in Zend.

In Gothic the symbol *þ* has been retained to express a sound resembling *th* in then.

In Lithuanian *sz* is pronounced *sh*; *w* stands for *v*, and *j* for *y*; *ą*, *ę* are nasalized vowels, *y* the hard *i*, as in Russian.

The exact differences in pronunciation denoted by the modified letters in Tocharian, *t*, *k*, *c*, etc., is uncertain; they correspond to special letters in the Tocharian texts; otherwise the orthography of Tocharian follows that adopted for Sanskrit.

THE ARYANS

NEW LIGHT ON INDO-EUROPEAN ORIGINS

CHAPTER I

LANGUAGE AND PREHISTORY

Man's progress from savagery to civilization is intimately bound up with the advance of abstract thinking, which enables him to rise above the chaos of particular sensations and to fashion therefrom an ordered cosmos. The growth of reasoning in its turn goes hand in hand with the development of language. The substratum of modern intellectual activities is very largely composed of those syntheses of audile and muscular sensations or images which represent words. These are not only means of communication, but also the vehicles of our abstract ideas. Words are the very stuff of thought. It follows then that a common language does imply a common mental outlook in its speakers; it not only reflects but also conditions ways of thinking peculiar to the users of the tongue in question. Moreover, intellectual progress may to a large extent be measured by the refinement of language. Hence to inherit an exceptionally delicate linguistic structure gives a people a vantage point on the path of progress.

Philology may therefore claim a place among the historical disciplines, the functions of which are to reanimate and interpret the process whereby man has raised himself from animalism to savagery, from savagery to barbarism, from barbarism to civilization. The painful steps of this advance at first lie beyond the reach of all written records. That is especially the case with the early cultures from which the contemporary civilization of the white races in Europe and in America is directly descended. Archæology, co-operating with anthropology, can indeed throw much light on the later phases of the process; it can provisionally identify the material forces under which certain types of culture have been generated and flourished, and the currents of trade and of migration which fostered their growth. But the individuality of the groups thus distinguished eludes explanation in abstract material terms. Why, for instance, had Europe, starting on the race 1,500 years behind Mesopotamia and Egypt, outstripped those pioneers in a millennium? Why did our continent then continue to progress while the Ancient East stagnated or declined? Favourable climatic conditions, peculiar natural resources, a happy conjuncture of trade routes do not suffice to explain this phenomenon; behind it lurks the true historic fact of personal initiative. That archæology cannot grasp, indeed the concrete person lies beyond the sphere of prehistory. But an approximation thereto in terms of racial individuality is attainable with the aid of philology. Language, albeit an abstraction, is yet a more subtle and pervasive criterion of individuality than the culture-group formed by comparing flints and potsherds or the "races" of the skull-measurer. And it is precisely in Europe, where the critical point of cultural evolution lies enshrouded in the gloom of the prehistoric period, that the linguistic principles just enunciated are most readily applicable.

Most of the languages of Europe, America, and India to-day belong to one linguistic family generally called the Indo-European. The direct ancestors of these modern tongues were already diffused from the Atlantic to the Ganges and the Tarim many centuries before our era opens; all seem to be descended from a common parent language (or, rather, group of dialects) which comparative philology can reconstruct in a schematic way. Naturally the parent language must have been spoken by actual people. These we shall call Aryans, and about them we can predicate two things.

To whatever physical race or races they belonged, they must have possessed a certain spiritual unity reflected in and conditioned by their community of speech. To their linguistic heirs they bequeathed, if not skull-types and bodily characteristics, at least something of this more subtle and more precious spiritual identity. Anyone who doubts this would do well to compare the dignified narrative carved by the Aryan Darius on the rock of Behistun with the bombastic and blatant self-glorification of the inscriptions of Ashurbanipal or Nebuchadrezzar.

Secondly the Indo-European languages and their assumed parent-speech have been throughout exceptionally delicate and flexible instruments of thought. They were almost unique, for instance, in possessing a substantive verb and at least a rudimentary machinery for building subordinate clauses that might express conceptual relations in a chain of ratiocination. It follows then that the Aryans must have been gifted with exceptional mental endowments, if not in enjoyment of a high material culture. This is more than mere inference. It is no accident that the first great advances towards abstract natural science were made by the Aryan

Greeks and the Hindus, not by the Babylonians or the Egyptians, despite their great material resources and their surprising progress in *techniques*—in astronomical observation for example. In the moralization of religion too Aryans have played a prominent rôle. The first great world religions which addressed their appeal to all men irrespective of race or nationality, Buddhism and Zoroastrianism, were the works of Aryans, propagated in Aryan speech.

It is quite possible that the Iranian Zoroaster anticipated even the Hebrew prophets in sublimating the idea of divinity, emancipating it from tribal or material trappings and enthroning an abstract righteousness where personified natural or magical forces had previously reigned. It is certain that the great concept of the Divine Law or Cosmic Order is associated with the first Aryan peoples who emerge upon the stage of history some 3,500 years ago (see p. 20 below). Even the original Aryans themselves worshipped at least one deity, a Sky Father,[1] who, although still anthropomorphic, materialistic and barbaric, was, nevertheless, exalted far above the nameless spirits and magic forces of mere savagery (see p. 81).

Nor were the potentialities of Aryan speech solely intellectual. Poetry in which a fixed metrical structure combines with sweet-sounding words to embody beautiful ideas seems peculiarly Aryan: Semitic poetry, for example, does not rest upon a regular metrical structure involving a fixed number of syllables in the verse. The correspondences between the metres of the Hindu Vedas, the Iranian Gathas, and the Greek lyrics, in fact, allow us to infer some form of common metrical tradition inherited from an earlier epoch.[2]

Thus philology reveals to us a folk whose language was pregnant with great possibilities. Now it was the linguistic heirs of this people who played the leading part in Europe from the dawn of history and in Western Asia during the last millennium before our era. It is perhaps then not overbold to hope that a collaboration between the two prehistoric disciplines of philology and archæology, at least in this modest domain, may help to solve certain problems that either science alone is powerless to resolve.

The Indo-European languages, when they first come within our ken in the middle of the 2nd millennium B.C., appear already

[1] Sans. *Dyauṣ pitā*, Gr. *Ζεύς*, Lat. *Juppiter*, Teut. *Tiu*.

[2] Meillet, *Les origines indo-européens des metresgrecques*, cf. Arnold, *Vedic Metres.*

dispersed in several distinct groups. The parent speech from which all are descended is itself preserved in no written documents, we can only reconstruct it approximately by comparative methods. Philologists to-day recognize eleven groups of languages descended from the Aryan root, each group embracing a plurality of languages and each language being in actual life divided up into a multiplicity of dialects. The principal groups known to-day are: (1) Celtic surviving only in Gaelic, Irish, Manx, Welsh, and Breton, but once spoken over a vast area in Western and Central Europe; (2) the Teutonic languages, including Anglo-Saxon, Dutch, German, and the Scandinavian languages, the oldest extant remains being a translation of the Gospels into Gothic by Ulfilas composed about 500 A.D.; (3) the Italic group—Latin, Oscan, and Umbrian all known from about 400 B.C.—together with their modern descendants, Italian, French, Spanish, Roumanian, etc.; (4) Albanian, possibly a survival of ancient Illyrian or Thracian; (5) Greek, in classical times divided into four groups of dialects; (6) the Slavonic tongues—Russian, Polish, Czech, Croat, Serbian, Bulgarian, and many others—the oldest monuments of which were written in Old Bulgarian or Church Slavonic about 900 A.D.; (7) the Baltic family Lithuanian, Old Prussian, and Lettic, all known only from a comparatively late epoch; (8) Armenian with a literature beginning in the sixth century A.D.; (9) Iranian dialects represented first in the Old Persian inscriptions of the Achaemenid kings on the one hand, and in the Gathas and later sacred books of the Parsis (Zend) on the other, and then in a great number of disparate dialects once diffused over an enormous area from Eastern Turkestan to the Caucasus and Europe (with the Alans) and still surviving in Ossetian, Kurdish, Persian, etc.; (10) Indic, primarily Sanskrit, then the ancient Prakrits, and finally the modern vernaculars; (11) Tocharian—an extinct language with two dialects known only from ancient manuscripts recently unearthed among the buried cities of the Tarim valley and probably dating from the later half of the 1st millennium A.D.

These eleven groups are doubtless only a fraction of the total number of Aryan languages which have once existed. The scanty fragments of ancient Phrygian, Messapian, and Venetic make it probable that these extinct tongues belonged to the Indo-European family. How many others there may have been which have vanished without leaving any trace we can only surmise. At the moment of writing quite unexpected traces of an Aryan language spoken in

Cappadocia during the 2nd millennium B.C. are coming to light. Are these eleven divisions final? Many endeavours have been made to simplify the scheme.

And, in fact, the eleven distinct families may be reduced to nine. The Baltic tongues, although more archaic, are so closely related in phonetics, structure, syntax, and vocabulary to the Slavonic that the two may be conveniently treated as a single group under the name of Balto-Slavonic. The same procedure can be applied with even greater security to the Indic and Iranian groups: the Sanskrit of the Rigveda and the Iranian of the inscriptions of Darius the Great and the Gathas of Zoroaster are so much alike that they might almost be regarded as just dialectic varieties of a common stock. Indeed, the connections of the Indians and the Iranians are not linguistic only. Both people called themselves by the common name of Āryas (Airya, Ariya), both had once known a common set of rivers and places (e.g. Sarasvati and Hara'uvatiš), worshipped the same deities (Mitrá, Aryamán, Nā́satyā, etc.), with psalms of the same metrical structure, and shared in the Soma sacrifice and other rites presided over by the same priests (*hótar-zoatar*, *Átharvan-athravan*).[1] Such correspondences allow us to conclude that the Indians and Iranians are, indeed, two branches of one and the same people who had lived together long after their separation from the parent stem.

No such thoroughgoing agreement links any one of the remaining nine groups to one of its neighbours rather than another. Nevertheless certain similarities in restricted spheres have been detected and proposed as tests of closer kinship. The most important steps in this direction have been taken in the department of phonetics, and phonetic changes, i.e. changes in pronunciation, do constitute a very fundamental feature of a language and may rest upon an ethnic basis. The most generally accepted division is based upon the treatment of the primitive gutturals, notably *k*. Indo-Iranian, Armenian, Balto-Slavonic, and, apparently, Thracian, all change *k* into a sibillant *s*, the remaining groups preserve the stop sound (which becomes *h* in Teutonic in accordance with Grimm's Law). The languages which change *k* to *s* further palatalize the sounds represented in Latin and Teutonic by labio-velars, q^u and g^u. The test word, which reveals the characteristics of the two divisions, is the name for the numeral 100. The *s* languages

[1] Cf. p. 33 below; the resemblances are conveniently summarized by Griswold, *The Religion of the Rigveda*, Oxford, 1924, pp. 21 ff.

are accordingly called *Satem* tongues and the others *Centum* languages.

The special significance formerly attached to this division was due to the belief that all the *satem* languages lay in Asia or the extreme east of Europe, while the *centum* groups would be restricted to the west. The treatment of the gutturals therefore seemed to mark a geographical as well as a linguistic division. This opinion has received a rude shock within the last decade; ancient manuscripts, written in an Indian alphabet discovered among the ruined cities of the Turfan and Khotan oases, proved to be composed in an Aryan tongue, Tocharian, belonging to the *centum* section! Thus *centum* speech was not confined to Europe; indeed, the easternmost Indo-European language spoken about the eighth century A.D. belongs to its side! Attempts have been made to escape the difficulty by proving that Tocharian was a comparatively late arrival in Chinese Turkestan, and was carried thither by a band of Celts.[1] It cannot be said that the efforts to connect Tocharian with Celtic rather than any other *centum* tongue have been crowned with any measure of success. Nevertheless, Tocharian does differ fundamentally not only in phonetics, but also in vocabulary and inflection from the other Aryan languages of Asia—Iranian and Indian—and moreover possesses a whole series of words which otherwise are peculiar to the European tongues.[2] Hence the discovery of Tocharian does not destroy the value of the division into *satem* and *centum* speeches, but only complicates its interpretation.

There are, of course, other phonetic peculiarites shared by more than one language or family. Thus Greek agrees with Iranian in changing pure initial or intervocal *s* into *h* in most cases. Again, most Celtic tongues, two Italic dialects (Oscan and Umbrian) and Aeolic Greek labialize *q*, but Old Irish and Gaelic, and two Italic dialects keep the guttural while the other Greek dialects only labialize it before *o* vowels using *t* before *e* and *i*.[3] Both these peculiarities then

[1] So Giles (C. A. H. ii). But though middle forms in -*ṭar* or *ṭr* and the 3rd plur. pret. in -*are* recall Celtic and Italic forms, the past participles in -*l* suggest just as close affinity with Slavonic. Pokorny (*I.J.*, 1924, p. 43) shows that Celtic peculiarities such as the assimilation of *p* to *q* before *q* (Lat. quinque, O. Ir. *cōic* as against Toch. *piś*) are missing in Tocharian and finds its closest analogies in Armenian phonetics. Finally some special affinities in vocabulary to Greek have been noted, e.g. *soyä* = *υἱος*, son.

[2] Such are *salyi*, salt, *laks*, fish (Ger. *lachs*, R. *lasoĭ*, salmon) and *alyek*, other. See Sieg and Siegling, *Tocharische Sprachreste*, 1921: Meillet, *I.J.*, 1913, and *C.Q.*, xviii, pp. 124 ff.

[3] E.g. W. *pimp*, O-U. *pumpe*, Aeol. *πεμπε* as against Ir. *coic*. Lat. quinque, Att. *πέντε*.

cut across the much more fundamental classifications, into Indo-Iranian and Greek in the one case and into Italic, Celtic, and Greek in the other. So they are not suited to act as bases of division. It must also be noted that the division into *satem* and *centum* tongues is phonetic only and could be crossed by other divisions founded on grammatical structure or vocabulary. Thus Greek (*centum*) and Sanskrit (*satem*) seem much more nearly allied in their verbal system than Sanskrit and Slavonic or Greek and Latin. Again, there is a very substantial number of words common to Indo-Iranian and Greek that do not recur in any other Indo-European tongues. On the other hand all the European languages, *centum* and *satem* alike, share a large vocabulary of terms which are strange to Indo-Iranian.

To clarify further our conception of the mutual relations of the nine Indo-European linguistic groups it may be well to dwell for a moment on the partly parallel case of the Romance languages. As everyone knows, French, Spanish, Portuguese, Catalan, Italian, Roumanian and the rest are descended from dialects of Latin—not the literary tongue of Livy or Tacitus, but the speech of the camp and the market place. This Low Latin as it is called must once have been intelligible from the Black Sea to the Atlantic. On the break up of the Roman Empire it gave place to a series of local dialects, each mutually intelligible to their immediate neighbours only.[1] Then political events or great authors raised certain of these dialects to be the official and literary languages of new realms—the dialects of North Castille and the Ile de France, for instance, became the regular media of communication throughout the kingdoms of Spain and France respectively. These State languages gradually ousted and suppressed the old gradation of dialects till to cross a political frontier meant to pass into the domain of an alien and unintelligible speech. Yet the national tongues spoken on either side of the border were equally derived from the common Low Latin substratum.

The linguistic divergences which now sundered the nations were due to phonetic change (i.e. differences in the pronunciation of the Latin sounds), innovations in inflection and syntax, and the adoption of diverse vocables whether variants existing in Latin itself, or distinct new formations from Latin roots, or derived from the pre-Roman languages of the province or again borrowed from later invaders and neighbours. But the divergences were of a regular order, and in the case of phonetics follow definite " laws " ; such phonetic modifications do not, however, affect words borrowed

[1] Vendryes, *Language*, pp. 264 f. Isaac Taylor, *Origin of the Aryans*, p. 264.

by the several languages after their constitution as separate entities, such as telephone or tobacco. "New" words can thus be distinguished from those inherited from the original Low Latin. By comparing the several distinct languages and applying the phonetic and other "laws" governing their differentiation so as to eliminate later borrowings, we could roughly reconstitute the parent language, even were no inscriptions or written monuments of it extant.

The divergences which mark off the several Indo-European groups of languages are of the same order as those operative in Romance. In the realm of phonetics the same absolute regularity is observable; changes in the pronunciation of the original Aryan words can be reduced to perfectly exact and universally valid "laws". In their light words descended from the parent speech can be at once distinguished from later borrowings by their conformity to such laws. Inflection has been affected by disintegration to a much greater degree than within the Romance group; the individual languages have in some cases equipped themselves with a whole mass of new formations, generally modelled on the old,[1] or have tended to simplify grammar by assimilating exceptional formations to more common types,[2] and later by replacing an inflectional by an analytic structure. These alterations have naturally entailed corresponding modifications in syntax. Finally the discrepancies in vocabulary are enormous, but that is not surprising; many of the Aryan languages have superseded older tongues, preserving from the latter many names for novel things or concepts, and the culture of the Aryans itself has been evolving very rapidly for centuries, necessitating the creation of new names. Still, as in the case of Romance, it remains possible to reconstruct the original Aryan speech in an abstract way by comparative methods.

Nevertheless, the analogy must not be pressed too far. To conceive of the parent speech as a mature language with a stereotyped vocabulary and rigid grammatical conventions like Latin would be grotesque. Such a fixed language only exists under the shelter of a stable and partly centralized political organization and enshrined in a written or traditional literature.[3] What we are

[1] Thus in Greek the passive, in Sanskrit the passive and the future, and in Latin all the tenses except present and perfect are new formations.

[2] For instance the replacement of reduplicated perfects or *s* aorists in Latin by perfects in *vi* or *ui*.

[3] Vendryes, *Language*, p. 261.

accustomed to call " languages "—the common language of modern newspapers, Greek historians or Babylonian legislators—could not exist in the social and material state of the primitive Aryans, as we shall describe it in Chapter IV. In fact, philology tells us that the parent language must really have been in a fluid state; from household to household, from generation to generation the pronunciation, inflection, and signification of words would vary a little. Comparison gives us only the abstract residuum when these historical diversities are ignored. Again, the causes which led to the diffusion of the Latin language—the creation of the Empire governed from Rome—presuppose a very exceptional degree of social organization and must not be taken as the type of linguistic diffusion.

Nevertheless, our analogy gives us further help. In the first place the extension of the Latin language in the Roman Empire presupposes the historic Roman people who created and spoke that language. On the other hand, their linguistic heirs, the speakers of Romance languages, belong neither historically nor anthropometrically to a single race. The bulk of the French and the Spaniards, for instance, is descended from various Ligurian, Iberian, and Celtic stocks who occupied Gaul and Hispania in pre-Roman times, mingled subsequently with different ingredients due to migrations and conquests by Goths, Alans, Normans, Burgundians, Moors and others and infiltrations of Gipsies, Jews, and suchlike uprooted stragglers. So we cannot argue from unity of language to unity of race. The point is so important that I venture to adduce another example to drive it home. The Bantu languages spoken over an immense area in Africa from the Great Lakes to the Cape are at least as closely allied as members of the Aryan family. Yet their speakers include representatives of the most divergent physical types.[1]

Again, the modern languages of France, Spain, and Roumania are not the result of a conquest or colonization of those regions by Frenchmen, Spaniards, or Roumanians. It was a single language that was spread, and that not so much by Roman settlement as by service in the legions and the convenience in law and commerce of the conqueror's speech to the Provincials themselves. The distinct languages have on the contrary evolved locally out of the linguistic continuum. To this extent the " undulation theory " propounded by J. Schmidt in 1872[2] and elaborated by Pictet and Isaac Taylor

[1] Johnston, *The Bantu Languages*, 1919, p. 25.
[2] *Die Verwandschaftverhältnisse der indogermanischen Sprachen.*

gives a more probable account of the diffusion of Aryan speech than the older "family tree" view. The earlier philologists had conceived of the dispersion of the Indo-European languages as the result of a stream of Aryan peoples flowing from a single circumscribed centre; the stream bifurcated as it advanced and each branch in turn divided in like manner, the separate branches corresponding to the ultimate groups of distinct languages. Schmidt, on the other hand, explained the differentiation of the Indo-European languages in terms of the propagation of various linguistic modifications from different centres within a vast continuum. The latter view evidently accords better with the actual affinities of the several Aryan languages as described on page 8, and with the inferences to be drawn from the Romance parallel.

Nevertheless, the older hypothesis must be invoked to explain the geographical position of some Aryan tongues. It is, for instance, very hard to account for the situation of Tocharian, an island of *centum* speech with marked European affinities in inflection and vocabulary, surrounded by a sea of *satem* Asiatic tongues, save by the assumption of an actual migration. At the same time the correspondences among the Aryan languages are far too close to allow the area of characterization of the parent speech to be regarded as illimitable. The Aryan cradle must have had a geographical unity; the linguistic data alone presuppose a block of allied dialects constituting a linguistic continuum within a specific area and under more or less uniform geographical conditions. The fact that the Aryans worshipped at least one common deity allows us to go further; for it implies not, indeed, political unity, but at least that the authors of the parent speech constituted a single people. To explain the distribution of Indo-European languages in prehistoric times we must then have recourse to some hypothesis of expansion, migration, conquest, or infiltration whereby Aryan speech and cult was carried from the "cradle land" to regions previously un-Aryan. To trace that expansion is the primary aim of this book. We shall first try to delimit the field of observation by locating the several Aryan peoples as they enter the stage of history. Then we shall seek to trace them back towards some common centre by the aid of archaeological remains. The counter part of this inductive study will be deductive. We shall endeavour to identify the primitive material culture and cradle of the Aryans as revealed by linguistic palaeontology among the cultural groups and provinces of the prehistorian. From this point it should be possible to trace the migrations of the Aryans to their historic seats by cultural evidence. The convergence of the two lines of research would test the validity of our conclusions.

APPENDIX

As illustrations of the relationships of the Indo-European languages, the following rough comparative tables may be of interest to the reader who is not a philologist :—

Numerals

	Sanskrit.	Greek.	Latin.	Irish.	Gothic.	Lithuanian.	Tocharian.
1	*éka*	εἷς	*unus*	*oen*	*ains*	*vė́nas*	*ṣom*
2	*dvā́*	δύω	*duo*	*dau*	*twai*	*dù*, F. *dvì*	—
3	*tráyah*	τρεῖς	*tres*	*tri*	**þreis*	*trỹs*	*trai*
4	*catvā́rah*	τέσσαρες	*quattuor*	*cethir*	*fidwor*	*keturì*	*śtwer*
5	*páñca*	πέντε	*quinque*	*coic*	*fimf*	*penkì*	*piś*
6	*ṣáṭ*	ἕξ	*sex*	*se*	*saihs*	*szezì*	*skas*
7	*saptá*	ἑπτά	*septem*	*secht n-*	*sibun*	*septynì*	*ṣukt*
8	*aṣṭaú*	ὀκτώ	*octo*	*ocht n-*	*ahtau*	*asztůnì*	*okt*
9	*náva*	ἐννέα	*novem*	*noi n-*	*niun*	*devyni*	*nu*
10	*dáśa*	δέκα	*decem*	*deich n-*	*taihun*	—	*śak*
100	*śatám*	ἑκατόν	*centum*	*cet*	*hund*	*szim̃tas*	*k̲ante*

VERB "TO BE"

Present

Sing.	Sanskrit.	Greek.	Latin.	Irish.	Gothic.	Lithuanian.	Armenian.
1st	*ásmi*	εἰμί	*sum*	*am*	*im*	*esmì*	*em*
2nd	*ási*	εἶ(ἐσσι)	*es*	*at*	*is*	? *esi*	*es*
3rd	*ásti*	ἐστί	*est*	*is*	*ist*	*ẽstì*	*e*
Dual.							
1st	*svás*	—	—	—	*siju*	*ẽsva*	—
2nd	*sthás*	ἐστόν	—	—	*sijuts*	*ẽsta*	—
3rd	*stás*	ἐστόν	—	—	—	—	—
Plur.							
1st	*smás*	εἰμές	*sumus*	*ammi*	*sijum*	*ẽsme*	*emk'*
2nd	*sthá*	ἐστέ	*estis*	*adib*	*sijuþ*	*ẽste*	*ēk'*
3rd	*sánti*	εἰσί	*sunt*	*it*	*sind*	(O.S. *sąt'*)	*en*

	Optative.				Imperfect.	
Sing.	Sanskrit.	Greek.	Latin.	Gothic.	Sanskrit.	Greek.
1st	*syā́m*	εἴην	*siem*	*sijau*	*ā́sam*	ἦα
2nd	*syā́s*	εἴης	*sies*	*sijais*	*ā́s*	ἦσθα
3rd	*syā́t*	εἴη	*siet*	*sijai*	*ā́s, āsīt*	ἦ
Plur.						
1st	*syā́ma*	εἶμεν	*simus*	*sijaima*	*ā́sma*	ἦμεν
2nd	*syā́ta*	εἶτε	*sitis*	*sijaiþ*	*ā́sta*	ἦτε
3rd	*syúr*	εἶεν	*sient*	*sijaina*	*ā́san*	ἦσαν

VERB "TO BEAR"

Present

Sing.	Sanskrit.	Greek.	Latin.	Irish.	Gothic.	Old Slav.	Armenian.
1st	*bhárāmi*	φέρω	*fero*	*-biur*	*baira*	*bera̧*	*berem*
2nd	*bhárasi*	φέρεις	*-is*	*biri*	*bairis*	*bereši*	*beres*
3rd	*bhárati*	φέρει	*-it*	*berid*	*bairiþ*	*beretʻ*	*bere*
Plur.							
1st	*bhárāmasi*	φέρομεν	*ferimus*	*bermai*	*báiram*	*beremʻ*	*beremkʻ*
2nd	*bháratha*	φέρετε	*-itis*	*-berid*	*bairiþ*	*berete*	*berekʻ*
3rd	*bháranti*	φέρουσι	*ferunt*	*berait*	*bairand*	*bera̧tʻ*	*beren*

Imperfect

		Sanskrit.	Greek.	Irish.	Armenian.
Sing.	1st	*ábharam*	ἔφερον	—	*beri*
	2nd	*ábharas*	ἔφερες	*-bir*	*berer*
	3rd	*ábharat*	ἔφερε	*beir*	*eber*
Plur.	1st	*ábharāma*	ἐφέρομεν	*-beram*	*berakʻ*
	2nd	*ábharata*	ἐφέρετε	*-berid*	*berekʻ*
	3rd	*ábharan*	ἔφερον	*berat*	*berin*

Middle Present

		Sanskrit.	Greek.	Gothic.
Sing.	1st	*bháre*	φέρομαι	—
	2nd	*bhárase*	φερει	*bairaza*
	3rd	*bhárate*	φέρεται	*bairada*
Plur.	1st	*bhárāmahe*	φερόμεθα	—
	2nd	*bháradhve*	φέρεσθε	—
	3rd	*bhárante*	φέρονται	*bairanda*

PERFECT VERB: "I HAVE COME TO KNOW"

		Sanskrit.	Greek.	Latin.	Gothic.
Sing.	1st	*véda*	οἶδα	*vidi*	*wait*
	2nd	*véttha*	οἶσθα	*vidisti*	*waist*
	3rd	*véda*	οἶδε	*vidit*	*wait*
Plur.	1st	*vidmá*	ἴδμεν	*vidimus*	*witum*
	2nd	*vidá*	ἴστε	*vidistis*	*wituþ*
	3rd	*vidúr*	ῒσασι	*videre*	*witun*

O-STEM NOUN "WOLF"

	Sanskrit.	Greek.	Latin.	Gothic.	Lithuanian.
Nom.	*vṛ́kaḥ*	λύκος	*lupus*	*wulfs*	*vil̃kas*
Acc.	*vṛ́kam*	λύκον	*lupum*	*wulf*	*vil̃ką*
Gen.	*vṛ́kasya*	λύκοιο	*lupi*	*wulfis*	*vil̃ko*
Dat.	*vṛ́kāya*	λύκῳ	*lupo*	(*wulfa*)	*vil̃kui*
Abl.	*vṛ́kāt*	*λύκω	*lupod*	—	*vil̃ko*
Loc.	*vṛ́ke*	*λύκοι	*lupi*	—	*vilkè*
Ins.	*vṛ́kena*	—	—	—	*vilkù*
Dual.					
N.-A.	*vṛ́kāu*	λύκω	—	—	*vilkù*
D.-I.	*vṛ́kābhyām*	—	—	—	*vilkam̃*

	Sanskrit.	Greek.	Latin.	Gothic.	Lithuanian.
	Plural.				
N.	*vṛ́kāḥ*	λύκοι	*lupi*	*wulfos*	*vilkaĩ*
A.	*vṛ́kān*	λύκους	*lupos*	*wulfans*	*vilkùs*
G.	*vṛ́kānām*	λύκων	*luporum*	*wulfe*	*vilkũ*
D.-A.	*vṛ́kebhyah*	—	—	*wulfam*	*vilkáms*
L.	*vṛ́keṣu*	λύκοισι	*lupis*	—	*vilkũsu*
I.	*vṛ́kaiḥ*	λύκοις	*lupis*	—	*vilkaĩs*

CONSONANTAL (N-) STEM

	Sanskrit.	Greek.	Latin.	Irish.	Gothic.	Lithuanian.
N.	*śvā́*	κύων	*homo*	*cu*	*hana*	*szů̃*
A.	*śvā́nam*	κύνα	*-inem*	*coin n-*	*hanan*	*szùnį*
G.	*śúnaḥ*	κυνός	*-inis*	*con*	*hanins*	*szuñs*
D.	*śúne*	—	*-ini*	*coin*	—	*szùniui*
L.	*śúni*	κυνί	*-ine*	—	*hanin*	*szunyjè*
	Plural.					
N.	*śvā́naḥ*	κύνες	*-ines*	*coin*	*hanans*	*szùns, szùnys*
A.	*śúnaḥ*	κύνας	*-ines*	*cona*	*hanans*	*szunìs*
G.	*śúnām*	κυνῶν	*-inum*	*con n-*	*hanane*	*szunũ*
D.	*śvábhyaḥ*	—	*-inibus*	*(conaib)*	*hanam*	*szunìms*
L.	*śvásu*	κυσί	—	—	—	*szunysè*
I.	*śvábhiḥ*	—	*-inibus*	*conaib*	*hanam*	*szunimìs*

DEMONSTRATIVE PRONOUN TO—

	Sanskrit.	Greek.	Latin.	Gothic.	Lithuanian.
N.	*sáḥ, tát, sā́*	ὁ τό ἡ	*iste, tud, ta*	*sa þata, so*	—
A.	*tám, tát, tā́m*	τόν τό τήν	*-tum, tud, tam*	*þana, þata, þo*	*tą̃ tą̃*
G.	*tásya, tásyāḥ*	τοῖο τῆς	*-tius*	*þis þizos*	*tõ tõs*
D.	*tásmai, tásyai*	τῷ τῇ	*-to -ti*	*þamma þizai*	*támui taĩ*
L.	*tásmin, tásyām*	—	—	—	*tami toje*
I.	*téna táyā*	τῶ	—	—	*tů́mi tà*
	Plural.				
N.	*té, tā́ni, tāḥ*	τοί τά ταί	*-ti, ta, tae*	*þai, þo, þos*	*tẽ, taĩ, tõs*
A.	*tā́n, tā́ni, tāḥ*	τούς τά τάς	*-tos, ta, tas*	*þans, þo, þos*	*tùs, taĩ, tàs*
G.	*téṣām, tā́sām*	τῶν τάων	*-torum, tarum*	*þize þizo*	*tũ tũ*
D.	*tébhyaḥ, tā́bhyaḥ*	—	—	*þaim þaim*	*tẽms tóms*
L.	*téṣu tā́su*	τοῖσι τῇσι	*-tis, tis*	—	*tůsè tosu*
I.	*taíḥ tā́bhiḥ*	τοῖς ταῖς	—	—	*taĩs tomìs*

CHAPTER II

THE FIRST APPEARANCE OF ARYANS ON THE STAGE OF HISTORY

1. *The Aryan Dynasts in Mesopotamia in the Fifteenth Century B.C.*

Aryan peoples first emerge from the gloom of prehistory on the northern borders of the Fertile Crescent of the Ancient East. The oldest Aryan names and words that have come down to us are inscribed upon cuneiform tablets from Babylonia, Egypt and Cappadocia. But these first historic Aryans appear as late intruders in a region illumined by the light of written documents from the end of the IVth millennium. In Mesopotamia and the adjoining countries they have invaded the domain hitherto occupied by peoples of different linguistic antecedents.[1]

From the dawn of history two non-Aryan races inhabited the Tigris-Euphrates valley—Sumerians and Semites. The former, though concentrated in southern Mesopotamia from the earliest times, have left certain monuments of their presence as far north as Assur on the middle Tigris, while Sumerian art products, if not Sumerians, penetrated even to Astrabad on the shores of the Caspian. Semites were inextricably mixed with the Sumerians in Babylonia, and occupied the western plains as far as the Syrian coast and Sinai. These two races jointly created the marvellous civilization of Mesopotamia, the monuments of which are known to us from the middle of the IVth millennium B.C. onwards. There they established great empires which diffused their culture throughout adjoining countries. Somewhere about 2700 B.C. the kings of Agade had extended their dominions to the shores of the Mediterranean and very probably to Cappadocia. In any case, soon after 2500 B.C. a substantial Semitic colony in close political and commercial relations with Assyria and Babylonia was established in the Halys valley in command of the trade routes that led to the Black Sea on the one hand and to the Aegean on the other.

[1] On this see Moret, *From Tribe to Empire*, part ii, chap. iii, and part iii, chap. i, ii, and iii; Delaporte, *Mesopotamia*, esp. p. 43; *Cambridge Ancient History*, i, esp. pp. 552 ff., and ii, pp. 13 f.

The Mesopotamian records and pictorial monuments also reveal to us other peoples inhabiting the adjacent highlands, none of whom seem to be Aryan in the IIIrd millennium. To the east lived the Elamites [1] speaking an agglutinative or incorporating tongue and possessed of a high civilization of their own. The highlands north of Irak were perhaps already occupied by Armenoid peoples, whose Asianic speech may be inferred from the later Hittite, Mitannian, and Vannic texts. Our sources give us no indication of the presence of Aryans within their purview down to 2000 B.C.[2]

FIG. 1. Naram-Sin, King of Agade, Semite.

But by the middle of the IInd millennium we find Aryan princes installed within the Fertile Crescent, heirs of the civilization created by Sumerian and Semite. The circumstances of their coming escape us; Hammurabi's dynasty, which had finally unified

[1] The only decipherable monuments of the Elamite language (Anzanite) date from a much later epoch, but the kings' names allow us to infer that it was in use also in the IIIrd millennium B.C.

[2] For Dr. Christian's view that the peoples of Gutium and Subartu were ruled by Aryans in the IIIrd millennium (*M.A.G.W.*, lv, p. 189) there is not a scrap of evidence. The names from this area are specifically non-Indo-European (cf. *C.A.H.*, i, pp. 421 and 452).

Mesopotamia under the hegemony of Babylon, fell about 1900 B.C.[1] and after its fall comes a dark age for which written records largely fail us. The precursors of the Aryan invaders may be found among the Kassites, who established a dynasty at Babylon about 1760 B.C. This people originally dwelt east of the Zagros Mountains whence they had begun to filter into Babylonia already in the time of Hammurabi. But as a whole they were not Aryans. Though they adopted the Babylonian language and culture, the local scribes have recorded the Kassite names for god, star, heaven, wind, man, foot, etc.; not one of these is in the least Indo-European. Moreover, the majority of the personal names of the period collected by Clay [2] suggest rather a kinship between the Kassites and the Asianic folk to the north-west. Yet in the names of their kings occur elements recalling Indo-Iranian deities—Šuriaš (Sun-god cf. Sans. *Surya*) Indaš (cf. Sans. *Indra*), Maruttaš (cf. Sans. *Marutaḥ*, storm-gods) and -bugaš (cf. Iran. *baga*, god). Moreover, these Kassites introduced the use of the horse for drawing chariots into the Ancient East and its later Babylonian name *ṣusu* seems to be derived from the Indo-Iranian form **asu̯a* (Sans. *aśva*). It is then highly probable that the Kassite invasion was due to the pressure of Aryan tribes on the highlands of Iran, and that its leaders were actually Aryan princes.

Three centuries later, when the diplomatic archives found at Tell el-Amarna cast such a flood of light on the affairs of Western Asia, we find a distinctively Aryan dynasty ruling among the Asianic Mitanni on the Upper Euphrates. These princes had good Aryan names—Šutarna, Dušratta, Artatama—and also worshipped Indo-Iranian deities. In 1907 Hugo Winckler [3] startled the learned world by identifying the names of four gods, already familiar from the Indian Veda, invoked as witnesses to a treaty signed in 1360 B.C. between the kings of Mitanni and the Hittites. The divine beings who are named together with other gods—ten Babylonian and four native Mitannian—are Indra (*in-da-ra*), Varuna (*u-ru-v-na* or *a-ru-na*), Mitra, and the Nāsatyā twins (*na-ša-at-ti-i-ia*). Quite recently another document emanating from Mitanni has turned up among the Hittite archives from Boghaz Keui.[4] It deals significantly enough with horse-breeding and contains a series of

[1] This date may have to be reduced by 100 years or more. See Delaporte, p. 18.
[2] *Yale Oriental Series*, i.
[3] *M.D.O.G.*, xxxv, p. 51.
[4] *Z.D.M.G.*, lxxvi, pp. 250 ff. (Forrer).

Aryan numerals—*aika* (1), *teras* (3), *panza* (5), *satta* (7), and *nāv* (9)—in expressions like *aikavartanna vasannasaya* (? " one round of the stadium "). Finally we know that there existed among the Mitanni at this time a class of warriors styled *marianna* which has suggested comparison with the Sanskrit *māryā*, young men, heroes.[1]

So it is clear enough that the dynasts installed on the Upper Euphrates by 1400 B.C. were Aryans, closely akin to those we meet in the Indus valley and later in Media and Persia. But their subjects were non-Aryan Asianics, and the rulers had adopted the native language and the Babylonian script for their official correspondence, and apparently acknowledged local gods besides their own. And the movement which had brought them to the Euphrates did not stop there. During the same period the Tell-el-Amarna tablets mention Aryan princes in Syria and Palestine too—Biridašwa of Yenoam, Šuwardata of Keilah, Yašdata of Taanach, Artamanya of Zir-Bashan and others.[2] These too were probably mere dynasts ruling over non-Aryan Semitic subjects.

These numerals and divine and personal names are the oldest actual specimens of any Aryan speech which we possess. The forms deserve special attention. They are already quite distinctly *satem* forms; in fact, they are very nearly pure Indic. Certainly they are much more nearly akin to Sanskrit than to any of the Iranian dialects that later constituted the western wing of the Indo-Iranian family. Thus among the deities *Nāsatya* is the Sanskrit form as opposed to the Zend *Naoṇhaitya* and all the four gods are prominent in the oldest Veda, while in the Iranian Avesta they have been degraded to secondary rank (Mithra), converted into demons (Indra) or renamed (Varuna=Ahura Mazda). The numerals are distinctively Indic not Iranian; *aika* is identical with the Sanskrit *eka*, while 'one' in Zend is *aeva*. So the *s* is preserved in *satta*, where it becomes *h* in Iranian (*hapta*) and the exact form is found, not indeed in Sanskrit, but in the Prakrits which were supposed to be post-Vedic.

Even the personal names look Indic rather than Iranian. Thus *Biridašwa* has been plausibly compared with the Sanskrit *Bṛhadaśva* (owning a great horse). If this be right the second element, *-ašwa*, horse, is in contrast to the Iranian form *aspa* seen in Old Persian and Zend (cf. *Jāmaspa* and *Vištāspa* = Hystaspes). On the other hand, the element *Arta-* in Mitannian and Palestinian

[1] Moret, *op. cit.*, p. 241.
[2] *C.A.H.*, ii, p. 331.

names has many parallels in the later Iranian onomasticon; the concept of divine order, *Ṛta*, embodied in it was indeed known to the Vedic poets, but it is rarely used as a component of personal names in India.

When we seek to define precisely to which branch of the Aryan stock these Mitanni princes belonged there is room for divergence of opinion. When the Mitanni deities' names were first published, Jacobi,[1] whom Pargiter[2] and Konow[3] still follow, definitely accepted them as Indian and ascribed their introduction into Mesopotamia to a body of Sanskrit-speaking peoples from the Punjab. To this Eduard Meyer[4] replied that philologists had long ago recognized that Indians and Iranians had lived together as one body and had worshipped these very deities in common before the Indians had occupied the Indus Valley. The Indian divine names and numerals would then belong to a branch of this Indo-Iranian people at a period before their differentiation, i.e. before the sound shifts distinctive of Iranian, *s* → *h*, *śv* → *sp*, etc., had become operative. Finally, Hüsing[5] agrees that the dynasts were Indians, but Indians on their way to India; for he holds that the scene of the Indo-Iranian period must be laid north of the Caucasus.

The decision between these three views must await a discussion of the later history of Indians and Iranians respectively. Two highly significant facts are secure: firstly, the cleavage into *centum* and *satem* languages goes back to the middle of the IInd millennium B.C.; secondly, that peoples later known to us only east of the Tigris at that date extended much further west.

2. *The Problem of the Hittites*

But not only were there Aryans of the *satem* branch in the Ancient East by the XVth century B.C.; the presence in the vicinity of peoples of the *centum* division is attested by cuneiform documents of the same epoch. If the discovery of Indic names in North Syria created astonishment in 1907, the revelation of a *centum* Indo-European element in the Hittite speech of Cappadocia ten years later provoked incredulity.

The Hittites had been long known from Egyptian and Babylonian

[1] *J.R.A.S.*, 1909, pp. 721 f.

[2] *Ancient Indian Historical Tradition.*

[3] *The Indian Gods of the Mitanni*, Publications of the Christiania Indian Institute, No. 1.

[4] *Sitzb. K. Preus. Akad. der Wiss.*, 1908.

[5] *M.A.G.W.*, xlvi, *Völkerschichten in altem Iran*, p. 210.

records, from Biblical traditions, and from their own enigmatic monuments. Before the war no one would have thought of connecting them with Aryans. Yet this was precisely the result to which the decipherment of the first substantial body of Hittite texts written in an intelligible script (cuneiform) led Professor Hrozny.[1] Rumours of his discoveries leapt political frontiers and even amid the din of battle aroused lively controversy. His conclusions were at first received with scepticism and it now appears that the solution of the problem is by no means so simple as he thought.

In the first place the material at our disposal, the tablets from the State archives of the monarchs of Boghaz-Keui, only refers to the Hittites of Cappadocia and justifies no conclusions with regard to other "Hittites", for instance, those of Carchemish in North Syria. And then it is now clear that even in Cappadocia a large number of languages were current simultaneously.[2] Some of these—Professor Forrer's Balaic, Ḫarric (Mitannian), and Proto-Hattic, which last has the best claim to be called by us, as it was by the native scribes themselves, "Hittite"—are quite definitely un-Aryan. It is otherwise with the dialect in which the majority of the texts are written, the language called by Forrer Kanesian, by the scribes *Našili*, "our language." It seems certainly to exhibit Indo-European influence.

At the moment we are on slippery ground; the number of texts published is not very large, the decipherment of the local cuneiform script offers many pitfalls, any judgment must be provisional. Yet some points have won fairly general assent. In the inflection of nouns, pronouns, and verbs Našili betrays most striking similarities to Indo-European.[3] Of the six cases in the nominal declension four

[1] *Die Sprache der Hettiter*, 1917.

[2] See Forrer, *M.D.O.G.*, lxi, and *Z.D.M.G.*, lxxvi; Sayce in *Anatolian Studies presented to Sir William Ramsay*, pp. 390 ff.; *J.A.O.S.*, 1921.

[3] To illustrate the point I quote the following forms given by Friedrich and Forrer in *Z.D.M.G.*, loc. cit.:—

	Neuter noun.	2nd Pers. pronoun.
Nom.	*u̯atar*	*zig*
Acc.	*u̯atar*	*tug, tuḳḳa*
Gen.	*u̯etenas*	*tuel*
Dat. L.	*u̯eteni*	—
Abl.	*u̯etenaz*	—
Ins.	*u̯etenit*	—
Plurals.		2nd Pers. pronoun.
Nom. (masc.	*-es*	*sumes*
Acc. noun)	*-us*	*sumās*
Gen.	—	*sumel*
D.-L.	*-as*	
Abl.	—	*sumedaz*

		Verb in *-nu.*	h conjugation.
Sing.	1st	*vahnumi*	—*ahhi*
	2nd	*vahnusi*	—*ti*
	3rd	*vahnuzi*	—*i*
Plur.	1st	*vahnuoeni*	
	2nd	*vahnuteni*	
	3rd	*vahnuanzi*	
Verb		*da-* to set	
Sing.	1st	*dahhi* (Pres.)	*dahhun*
	2nd	*datti*	(Imperf.)
	3rd	*dai*	*dās*
	1st	—	—
	3rd	*danzi*	*dair*

admit of a plausible explanation from Aryan paradigms. With the pronouns the proportion is rather less. In one conjugation five out of six forms in the present and three in the past look Indo-European; to these may be added the middle endings in *-tari* and *-antari* and imperatives in *-du* and *-andu*. Some forms such as the pronouns *kuis*, *kuit* (cf. *quis*, *quid*), the verb *ešmi*, I am (cf. Sans. *asmi*), or again the formation of present stems in *-numi* (Greek *-νυμι*, Sans. *-nomi*) look extraordinarily Aryan.

But the most surprising thing about them is that the Indo-European resemblances lie not at all with Indo-Iranian, but with the *centum* languages, especially Phrygian, Greek and Latin. The phonetic system would connect Našili exclusively with the latter group. In inflection some forms have peculiarly clear western affinities: the accusative singular in *-n*, as in Greek and Phrygian, instead of *-m*, *sumes*, you like the Greek *ὑμεῖς*, the adverb *kattā*, and the change of *-ti* to *-zi* in the 3rd sing. of verbs. Only the rather dubious imperatives in *-du* and *-andu* and the 2nd Person Plurals in *-teni* have distinctively Indo-Iranian parallels. If there be an Indo-European element in Našili, it cannot be derived from their neighbours in Mitanni.[1]

However, Našili cannot be accepted without qualification as Aryan. The deviations in the inflection are puzzlingly numerous. Professor Sayce tells me that the very Indo-European looking endings of the verbal stem are not quite strictly "personal", but seem sometimes to be used indifferently to denote the first or third person, the singular or plural. And as he has pointed out several of the supposedly Indo-European verbal terminations, have parallels in non-Aryan languages, Vannic, and even Sumerian.[2] Again the number of Indo-European words and stems identified in the vocabulary is but small.[3] Finally, the syntax remains essentially un-Aryan, for the structure is "incorporating" as in the Asianic tongues.[4]

Now if these documents dated from the XIVth century A.D. few would hesitate to declare that they were written in an Indo-European language and explain the discrepancies as due to the familiar phenomena of decay, assimilation of forms, and foreign borrowing. But the texts from Boghaz-Keui are many centuries older than the

[1] Nor from Iranian Medes (Manda), as Giles suggests, *C.A.H.*, ii, p. 15.

[2] *J.R.A.S.*, 1920, p. 58.

[3] Many derivations have been proposed which, while plausible in themselves, taken together assume mutually incompatible phonetic laws.

[4] *Ramsay Studies*, p. 392.

earliest written memorials of Sanskrit or Greek. Yet their language diverges from the hypothetical original Aryan tongue far more than Greek and Sanskrit differ from the parent speech or one another. It is in fact impossible to believe that a truly Indo-European language would look so odd in the XIVth century before our era. Professor Forrer has suggested the possibility that Našili might be a branch of some very archaic tongue from which the parent Aryan speech was also sprung. I believe rather that the clue lies in recognizing with Professor Sayce that Našili was an artificial literary language elaborated by court scribes and priests.[1] In such a composite Aryan elements, words and terminations might be borrowed to express concepts and relations unknown to the more primitive Asianic dialects which constitute the substratum of the language. In the same way a whole mass of Babylonian terms have been incorporated. In a like manner the scribe of the Elamite version of the inscription of Darius at Behistun has adopted the Old Persian imperative *aštu* [2] since the substantive verb was missing in Anzanite.

If we then admit the real presence of an Indo-European element in the language of Cappadocia, we have still to ask whence it came. The usual answer is that the Aryan element, there as in Mitanni, was just the ruling aristocracy who had imposed themselves on an older Asianic substratum. However, the names of the Hittite kings—Hattusil, Dudḫalia, Mursil, Mutallu—do not look in the least Aryan. Again, no Hittite deities have Aryan names, though Professor Sayce has pointed out that in the Hittite version of the Babylonian myth of Bêl and the dragon, the monster has an Indo-European name—Illuyankas.[3] What a contrast to the Mitanni princes who kept their Aryan names and gods! Again the dynastic lists are said to take the dynasty with the same non-Aryan names back to 1900 B.C. if not earlier. It looks as if the kings of Boghaz-Keui belonged rather to an Asianic stock.

Moreover, this Asianic element can be traced back to the middle of the IIIrd millennium B.C. in Cappadocia. At that time colonies of Semites were established in the Halys valley, and it was doubtless from them that the Babylonian elements in Hittite culture and in the Našili language were borrowed. The correspondence of these

[1] Cf. Luckenbill in *J.R.A.S. Centenary Volume*, p. 58, who adduces interesting parallels from America.

[2] Col. iii, l. 65, as an equivalent of the subjunctive *ahatiy* in the Persian text, col. iv, l. 39.

[3] *J.R.A.S.*, 1922, p. 185: *illu* = Babylonian *ilu*, god, but *yankas* = ἔχις = *anguis* = Sans. *ahi*.

merchants, the so-called Cappadocian tablets, reveals the presence of people with Asianic names like those of the Hittite kings, Dudḫalia, Buzua, Aḫukar, etc., and contains references to Buruš Ḫatim, 'the Hittite fortress' before 2000 B.C.[1] A "Cappadocian" seal of about the same date is said further to bear a legend in Hittite hieroglyphics.[2] But no Aryan names occur so early. Whatever element in the population inspired the Aryanization of Našili, then, and wherever it dwelt, it looks as if it had only begun to influence the Halys region after 2000 B.C., perhaps long after. The only certain result that has emerged as yet is that there was a *centum* element somewhere within the Hittite realm just after 1500 B.C. About that date the Taurus ranges seem to have represented in a sense a frontier between *satem* and *centum* Indo-European speech.

3. *Archæological Pointers*

Whence then came these two groups of Aryan peoples appearing on either side of the Taurus? That they were intruders may be inferred from the silence of the documents of the IIIrd millennium. They are first definitely revealed to us at the end of a dark age. The darkness is itself significant; for it reflects the consequences of social convulsions provoked by the movement of peoples. Fresh ethnic elements had broken their way into Hither Asia. With their intrusion we may connect the invasion of Egypt by the Hyksos or Shepherd Kings. The documents of the XVth century allow us to infer the catastrophe and reveal the new alignment of political forces it created. But only archæology is likely to disclose the direction of the antecedent racial movements.

At the moment, unhappily, its contribution is small. The earlier phases of the Kassite period in Babylonia, before the invaders had completely assimilated the culture of their adopted country, are little known. It is nevertheless of interest to note that under Ammizaduga, just before the Kassite conquest, white slaves from Subartu and Gutium, regions to the north-east, were being sold at Babylon. About the same time the importation of jade from Chinese Turkestan seems to have been interrupted.[3]

[1] Sidney Smith, *Cappadocian Tablets in the British Museum*, Sayce in *J.H.S.*, xliii, pp. 44 f.

[2] Sayce in *J.R.A.S.*, 1922, p. 266.

[3] Kennedy, *J.R.A.S.*, 1909, pp. 1113 ff.

The Mitannians are even less known. Only one site in their territory has yet been explored—Tell el-Halaf on the Habur.[1] Here Baron von Oppenheim has excavated the ruins of an ancient city, and some of his finds are now in the British Museum. The rude bas-reliefs, in Hittite style, and seemingly depicting Hittite racial types, belong to the first millennium before our era, and so

FIG. 2. Asiatic enemies depicted on the chariot of Thothmes IV: 1, Naharaina; 2, Sangari; 3, Shasu (Beduin); 4, Kadshi.

throw no light on our immediate problem. The small objects, including painted pottery, from the lower strata may prove more enlightening when they are published. Nor have we any certain portraits of Mitannians. Yet the Aryan dynasty was in constant

[1] Cf. Oppenheim, *Der Tel Halaf, Der alte Orient*, x, 1908.

relations with the Pharaohs and no less than three Mitannian princesses became queens of Egypt. Of these, Mutemua, the wife of Thothmes IV, looks thoroughly Egyptian on her portrait statues [1]; that might be court convention. No portrait of the other two princesses, Gilukhipa and Tadukhipa, who entered the harem of Amenhotep III have come down to us. But earlier in the XVIIIth Dynasty the Pharaohs had been at war with these Aryans of North Syria. They have left us a regular portrait gallery of the races they had subdued in the course of their northern expeditions. The majority of these are certainly Semites—Amorites, Beduins, and so on—but the man from "Naharaina" on the chariot of Thothmes IV,[2] Fig 2, 1, seems to stand out from among the rest and recalls Iranian types from the Persian monuments a thousand years later. He may be one of the *marianna* since Naharina adjoins the Mitanni territory and Mitannians were actually met there by the conquering Pharaoh. It is also interesting that on Egyptian monuments of the XVIIIth and subsequent Dynasties the Amorites are often depicted as tall, relatively fair, blue-eyed, and brown-haired.[3] In view of the considerable Aryan infusion in the area inhabited by them it is just possible that this was not a native characteristic but was due to admixture with the intruders. A systematic exploration of the Mitannian territory and of the towns in Syria and Palestine where Aryan princes were installed should throw much further light on these problems. Pending such work I can only draw attention to certain phenomena which seem to mark innovations at the period of Aryan intrusion on the chance that they may serve as pointers.

Eduard Meyer [4] has called attention to a curious chariot, now in Florence, found in an XVIIIth Dynasty grave in Egypt. It is of a foreign type, and the axel is bound with birch-bark. Meyer says that that tree does not grow nearer than the Caucasus and accordingly suggests that the Aryans entered Hither Asia across those mountains like the Cimmerians and Scythians a thousand years later. Of course, the attribution of this particular chariot

[1] One in the British Museum; cf. Petrie, *History of Egypt*, ii, p. 173, fig. 111. Her Mitannian origin has been denied by several Egyptologists, most recently by Pridek in *Acta et Comment. Univ. Dorpatensis*, V.B.; cf. Moret, p. 291.

[2] Carter and Newberry, *Cairo Museum Catalogue, Tomb of Thutmôsis IV*.

[3] Clay, *The Empire of the Amorites*, p. 59. On the other hand, Prof. Sayce holds that the Amorites as such were a fair and blue-eyed stock related to the blonde "Libyans" and the Celts (*J.R.A.S.*, 1924, p. 115). Sir Flinders Petrie seems to favour a similar view.

[4] *M.D.O.G.*, lvii, p. 16.

to the Aryans is sheer speculation, though we have seen reason to believe that they did introduce the use of horse-drawn vehicles into the Ancient East. Be that as it may, there are other evidences of intercourse with the Caucasus region at about the same epoch.

The oldest cist-graves from the Carchemish region in North Syria contain objects paralleled in South Russia. These are eyelet-pins which look like an elaboration of types found in Copper Age barrows on the Kuban River, curious poker-butted spear-heads which also recur north of the Caucasus and penanular bracelets with flattened recoiled ends widely diffused in South Russia, Hungary, Upper Italy and Bohemia.[1] Of course, we have no sure grounds for regarding these European objects as older than their Syrian parallels nor yet for connecting the latter with Aryans whether "Hittitic" or Indo-Iranic. Still they may be pointers and point across the Caucasus.

On the other hand proofs of influence from Upper Asia may only be lacking because that region is still unexplored. As a matter of fact we do find traces of connection with Turkestan somewhere about this period, though they seem to be from the west and not *vice versa* (see page 40 below).

Passing to the Hittite area in Cappadocia we are slightly better documented.[2] Hittites are portrayed by their own sculptors and by Egyptian painters. The representations of these people from 3,000 years ago would serve as portraits of the Armenians who inhabit the same region to-day. They reveal a short-headed, high-skulled race, with a large nose and retreating forehead (Pl. I), which had then, as now, a wide extension in the highlands of Western Asia. Indeed, von Luschan [3] could call the modern Persians "Hittites". That does not, however, prove that this Armenoid race was the original Aryan stock either in Cappadocia or Iran; the Aryan Darius with his high-forehead and fine nose is quite different and is evidently Mediterranean or Nordic. The best eastern parallels to the Cappadocian Armenoids are to be found in figures of the Elamite goddess Anahita just as non-Aryan Kassite names find echoes in the Hittite territory. It looks therefore as if the typical Armenoid Hittite represents a pre-Aryan Alpine stock.

[1] See *L.A.A.A.*, vi, Woolley, *The Graves of the Hittites*; and Childe, *Dawn*, p. 146, and figs. 62 and 91, 11–12. Cf. *Syria*, vi, pp. 16 ff. (Byblos).

[2] Cf. Moret, *op. cit.*, p. 237; Garstang, *The Land of the Hittites*; Cowley, *The Hittites*; and Ed. Meyer, *Reich und Kultur der Chetiter*.

[3] *J.R.A.I.*, xli, pp. 242 f.

Even the Egyptian drawings of Hittite troops show other less Asianic types.

The typical articles of Hittite costume were the high boots with upturned toes and the peaked cap. The latter meets us again among the nomads of the steppes in both Persian and Greco-Scythian art (Pl. II, 2), and is still worn by the Mongols of Upper Asia. The pigtails worn by the Hittites again look Mongolian to Garstang but perhaps have analogies also in Syrian and Minoan coiffures. None of these peculiarities can on other grounds be regarded as specifically Aryan.

In culture the Hittites of Boghaz-Keui were of course indebted to Babylonia, but they were no slavish imitators. They did not, for instance, like the Assyrians after 2400 B.C., virtually abandon the use of stone in their buildings for the Babylonian brick architecture. The walls of Boghaz-Keui [1] indeed in the use of Cyclopean masonry recall the prehistoric acropoles of the Aegean. The megalithic orthostatae of the gates again have parallels at Troy, Tiryns, and Mycenae. But the gates themselves preserve the double straight Babylonian type and lack the flanking bastion introduced further west already during the habitation of Troy II. Even the plan of the palaces or temples within these walls with their system of cell-like basements built round a central court has been compared to the palace of Knossos in Crete.[2] But we cannot tell by inspection how far these Aegean parallels in Cappadocia are not a common " Asianic " heritage of the two regions.

It is moreover evident that contact between the Aegean and Cappadocia goes back to the IIIrd millennium. From Kara Euyuk, not far from Boghaz-Keui, come spouted vases of typical Early Minoan form and clay stamps with exact analogues in Troy II and Bulgaria. But similar vases are known from Persia and, accompanied by clay stamps of the Cappadocian pattern and spiral-headed pins of Aegean type, in the third settlement of Anau in Turkestan.[3] All this suggests trade rather than migration; its direction is still dubious, and its antiquity would seem to connect it with the pre-Aryan period in Cappadocia.

For the epoch and area for which alone we are warranted in speaking of an Aryan element among the Hittites we have only the

[1] Puchstein, *Boghazköi, Die Bauwerkerke* (*Wiss. Veröffentl. D.O.G.*, 46).

[2] But early houses in Sumer were similarly laid out. Delaporte, *Mesopotamia*, p. 111.

[3] Childe, *op. cit.*, pp. 26, 45. *Vide infra*, p. 111.

PLATE I

HITTITE DEITY

[face p. 28

carved monuments to guide us pending the publication of the small finds from Boghaz-Keui. The most characteristic weapon is the battle-axe borne by the god Teshub; actual specimens are known from Elam and Transcaucasia,[1] but the type is only an elaboration of a very ancient Mesopotamian weapon. On the other hand, the same deity carries on his left side a sword which is neither a Babylonian nor Egyptian weapon (Pl. I). It must be half a metre long and the blade looks so wide that it may be a slashing weapon and not, like all earlier oriental swords, designed only for thrusting. East of the Aegean these are certainly the longest weapons known from the IInd millennium. However, the hilt, which leaves a well-marked semicircular indent where it meets the blade, recalls Central European rather than Aegean types. It is therefore worth noting that a sword with a similar hilt and unmistakable Scandinavian

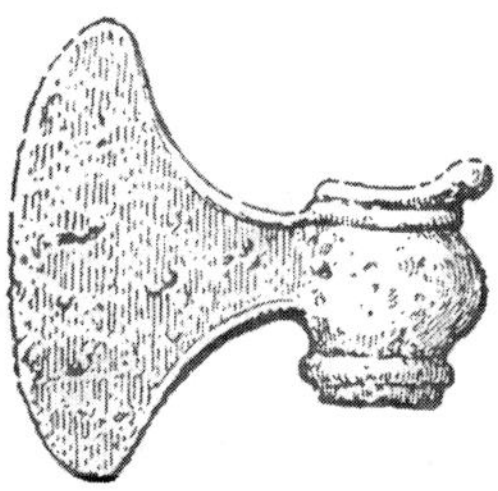

FIG. 3. Axe from Susa.

and Danubian affinities has been found at Mouçi Yeri in Armenia.[2] Moreover the girdle worn by the "Amazon" from Boghaz-Keui is strikingly similar in form to bronze girdles found in graves on both sides of the Caucasus, as Cowley remarks.[3]

Another Hittite weapon with a curving point has parallels as far west as Troy. The Hittite shield again resembles a diminutive version of the Mycenaean figure 8 type. Like the Kassites and Mitannians, the Hittites fought from chariots drawn by horses. This animal appears very early on Cappadocian seals. It should also be noted that the Hittites were ahead of all their neighbours in the industrial use of iron in the XIVth century.

The unambiguous links that shall bind the Aryan element in Cappadocia and North Syria during the IInd millennium indissolubly

[1] de Morgan, *Prehistoric Man*, fig. 54, i; *P.Z.*, iv, p. 32, fig. 3; Fig. 3 here.
[2] de Morgan, *op. cit.*, figs. 66, 7; Figs. 25, 7 here.
[3] *The Hittites*, fig. 10.

to some other area, be it the Caucasus, the Aegean, or Central Asia, are not yet forthcoming. The Cappadocian Aryans, like those of Mitanni, therefore remain for the moment isolated intruders who might have come from almost any quarter save the south. We have next to see if we can trace their later history in the hope that thence we may derive some light on the problems of origins.

4. *The Aryan Invasion of India*

In Palestine the Aryan names have totally disappeared by 1000 B.C., and even in the Mitanni region they leave scarcely a vestige behind them. Here at least Aryan speech succumbed to Semitic and Asianic dialects, and the small Aryan aristocracies were absorbed by the native population. Further east on the tablelands of Iran and in India the Aryan languages survived and survive to-day. But here written evidence still fails us till the VIth century B.C. Our oldest sources are the metrical compositions of the Hindus and the Iranians themselves, handed down for many centuries by oral tradition.

The Indians' language approximates most closely to that of the Mitanni documents and has been preserved from a remote date in the hymns of the Rigveda. This priceless document also furnishes precious historical data. The oldest Veda is a collection of metrical chants, always spirited, sometimes truly poetic, more rarely solemn and exalted. Their interest is naturally mainly religious. The powers of nature, sky and sun-gods, the lords of the storm and the wind, the dawn maiden and the heavenly twins, the sacred fire and the ritual intoxicant Soma are invoked in many a stanza. Dearest of all is the rollicking war-god Indra, the thunderer, made in the image of an Aryan chieftain of the heroic age. Exhilarated by Soma drafts, he slays the dragon Vṛtra or Ahi, releases the stolen kine of light or rain, and rescues the imprisoned Dawn. Only incidentally do we see the earthly princes whom Indra copies—generous to bards, bold to smite the dark-skinned Dasyus (aborigines), lovers of strong drink, dicing and horse-racing—in a word, with all the characters of a Teutonic hero in the Norse epic. Their wealth was in kine and horses, their vehicle the horse-drawn chariot, their weapons the bow, the mace and the spear. Axes of copper are mentioned, but as tools not weapons.[1] Temples or cities are not

[1] *paraśuh suayasá.* Macdonnell takes *ayas* to mean iron, but copper is intrinsically more likely and accords better with similes such as *ayodaṃṣṭra,* " with teeth of *ayas,*" applied to the fire-god. Cf. Zimmer, *Altindisches Leben.*

described, though strong places (*puraḥ*) are referred to. The dead are generally cremated, the ashes interred under a barrow.

This seems the picture of a young and vigorous race fresh from the mountains taking possession of the torrid plains of northern India. On the orthodox view they are still in the Punjab, whither they have but recently descended from Afghanistan. Rivers west of the Indus [1] and the district of Gandhara are still within the ken of the singers. The Jumna (Yamunā) is named but thrice, the Ganges twice, in seemingly late verses, but the oft mentioned Sarayu may be the River of Oudh (now Sarju). South India is quite unknown. It is accordingly believed that the centre of Aryan India in Vedic times lay in the Punjab. The eastward expansion would then belong to the later period of the Brahmanas. These are liturgical and epexegetical texts composed largely in prose and shown by their altered language, social ideas and religious outlook, to be separated from the Rigveda by a considerable interval of time. They stand to the Veda rather in the same relation as Hesiod to Homer. Yet even they are very distinctly pre-Buddhist and antedate the conquest of southern India. The Rigveda is therefore dated somewhere after 1400 B.C., and the Aryan invasion is assigned to a like antiquity.

This orthodox view has recently been challenged from two sides. Mr. Pargiter [2] holds that the Aryanization of India was long prior to the composition of the Vedic hymns. He complains that the usual European view has relied too implicity on the traditions of the Brahman caste. But there exists another tradition, that of the kingly or *kṣatriya* class. To this Mr. Pargiter appeals, though he admits that the existing redactions thereof are late.[3] A study of the royal genealogies leads our author to the conclusion that the Aryans, identified by him with the Aila or Lunar race, entered India nearer the beginning than the end of the second millennium and over the Central Himalayas, not across the Hindu Kush. Their oldest centre was, on this view, on the Upper Jumna and the Ganges; the occupation of the Punjab and the Rigveda belong to a later age of westward expansion. Then, following Jacobi's interpretation of the Mitanni names, Mr. Pargiter assumes that the same wave spread still further west—into Mesopotamia.

If this account of the Aryanization of India be accepted, the whole

[1] The Kabul (*Kubhā*), Kurrum (*Krumu*), and Gomal (*Gomati*).
[2] *Ancient Indian Historical Tradition*, Oxford, 1922.
[3] See a good critique in *J.A.O.S.*, 1923, pp. 123 ff.

problem of this book will assume a new aspect. But geographically the transit of the Himalayas offers severe obstacles and other difficulties are entailed in the Iranian connections. And the *kṣatriya* tradition on which the whole theory is based is hardly an unpolluted source of history. The orthodox view is not really based on the priestly tradition, as embodied in epexegetical works, but rather on the internal evidence of the Veda itself. The latter carries conviction precisely because the historical and geographical references in the hymns are introduced only incidentally and in a thoroughly ingenuous manner ; for instance, there is no caste in the Rigveda, and the priest is obviously dependent upon the generosity of his kingly patron. The same cannot be said of the *kṣatriya* tradition, which in its recorded form dates from an age (perhaps as late as 200 A.D.) when myth-making had had many centuries to work in, and which might serve dynastic ends. It needs even more cautious handling than the tales of Homeric heroes in late logographers and Roman poets. So the traditional view is still perhaps the more convincing.

The latter has however been challenged from the contrary standpoint in recent years. Brunnhofer [1] and others have argued that the scene of the Rigveda is laid, not in the Punjab, but in Afghanistan or Iran, and this view has lately been espoused by Hüsing.[2] In that case the occupation of India would be much later than is usually assumed. Now Brunnhofer relies mainly on the identification of peoples mentioned in the Veda, with tribes located in Afghanistan, in the inscriptions of Darius, or in later Greek authors.[3] But his heretical views have not received much support among Indianists. Some of his identifications are indeed phonetically preposterous, but there is a residue which demands explanation ; the mention of *Parthāva* in the Veda is an old crux owing to the formal identity of the word with the Old Persian form of Parthian and the occurrence of *Parśu* with, or as an adjective of, such an Iranian sounding name as Tirindira (in *R.V.*, viii, 6, 46) makes the translation "Persian" very tempting. Moreover, there is not the least doubt about the phonetic equation of the Vedic Sarasvati with the Persian Hara'uvtiš and the Rāsā with the Zend Raṅha.

[1] *Arische Urzeit*, 1910.
[2] *M.A.G.W.*, xlvi.
[3] Thus Paṇi = *Πάρνοι*, Sṛñjaya = *Ζάραγγαι*, Mṛdha = *Μάρδοι*, Ānava = *'Αναβων, χώρα* (part of *'Αρεια*), Śiva = *Σιβοι*, etc.

In fact, it is beyond dispute that some transference of river names between Iran and India has taken place. But most of the other streams named in the Rigveda are Indian rivers which bore the same names in historical times.[1] To suppose with Brunnhofer that all these names had been transferred from the miserable streamlets of Seistan to India involves a far greater dislocation of topographical nomenclature than is required on the orthodox theses. This admits the nominal identity of Sarasvati and Hara'uvatiš, but supposes that the Vedic Hindus applied the name of a stream once known to them in company with the Iranians west of the Hindu Kush to the chief river of their new home, the Indus (*Sindhu*=the River), or to the modern Sarasvati, a small stream east of the Sutlej which now loses itself in sand.

The element of truth underlying the second heresy would then be the reality of the Indo-Iranian period which we have postulated on other grounds (page 7) and which adequately explains the phenomena just noted. But the same truth powerfully reinforces the orthodox theory against Mr. Pargiter too. He might indeed accept the curious view of Darmsteter [2] to this extent and regard the Vedic gods and rites known to the authors of the Avesta as borrowed from Indians who on his view overflowed from the Punjab in the XVth century B.C. Would he also ascribe the Aryanization of Iran to the same migrants ? In that case he would have to assume a numerically large band. But we have few, if any, examples of such a large scale emigration out of India ; the movement of peoples in historic times has generally been into that land. At the date in question the Aryans had still all southern India to colonize. Why then should they climb the passes of Afghanistan to wander on the bleak tablelands of Iran ? Moreover, the common myth of the dragon fight—Indra and Ahi in the Veda and Atar and Azi in the Avesta—seems at home in Mesopotamia. The coexistence of Indians and Iranians somewhere within the sphere of Babylonian influence would make its appearance in an Aryanized version on both sides of the Hindu Kush intelligible.

The weight of evidence then compels us to regard the coexistence of Indians and Iranians as pre-Vedic and to locate its scene west of the Hindu Kush. That implies an invasion from the west such as we have sketched on page 31.

[1] Kubhā = Kabul, Krumū = Kurrum, Gomati = Gomal, Śutudrī = Sutlej, Vipaś = Biyas, Paruṣṇī = Ravi, Aśiknī = 'Ακεσίνης, etc.

[2] *Sacred Books of the East*, *The Zand-Avesta*, vol. i, introduction. This author's date for Zoroaster is quite untenable.

Archaeological documents illustrating this invasion would be of quite exceptional value. But till 1924 scarcely any pre-Buddhist remains were known in northern India ; in the south, indeed, and in Assam stone circles and megalithic tombs containing cremated remains and iron objects have long been known, but these districts were only brought under Aryan rule at a late date and are still essentially un-Aryan. The material there revealed therefore has no bearing on our question. But last year traces of an entirely new culture going back to a chalcolithic epoch came to light in the Indus valley, in Sindh near Larkana, and in the Montgomery District of the Punjab just north of the Sutlej (Śutudri). These astounding discoveries are at present only known from summary reports,[1] but they do reveal unmistakable evidence of connection one way or the other with the west and that at a very remote epoch.

The civilization here laid bare undoubtedly lasted a long time, as several strata of ruins have been discovered. Some of the material, presumably the oldest, evinces obvious parallelism to early Mesopotamian remains ; the use of brick for building, the interments of contracted bodies in brick cist graves, the shell inlays, the mace-heads and pestles all have the most exact analogues in early Sumerian levels in the Tigris-Euphrates valley. The beautiful stamp-seals engraved with figures of *Bos primigenius* and (?) unicorns and the curious symbols of their legends likewise have good Sumerian counterparts, and so, to a less striking degree, have the clay models of rams and the female figurines. Finally the painted pottery from the Indus sites is connected through Baluchistan with Elam and Southern Mesopotamia and more vaguely with Seistan and Transcaspia.

Here we have for the first time positive evidence of intercourse between India and Western Asia before the first millennium—and these connections were evidently very ancient, presumably anterior to the general adoption of the cylinder seal in Mesopotamia about 2800 B.C. But at a later period in the history of the ruins a significant change took place in the civilization of the Punjab ; inhumation gave place to cremation.

The data available seem susceptible of three interpretations : either the whole civilization of the Punjab is Aryan, or the Aryan element enters at some date within the long ages represented by the accumulated debris—perhaps with the introduction of cremation —or finally the Aryans were just the destroyers of the newly

[1] *Illustrated London News*, 20th September, 1924. Cf. Pl. VI here.

discovered culture. We shall return to the first possibility in a later chapter, but here some preliminary points must be noted. The connections with Sumer and Elam in themselves suggest that the authors of this civilization were not Aryans but connected with one of the pre-Aryan races of Mesopotamia. Indeed, Dr. Hall [1] pointed out ten years ago that the Dravidians of India resemble in anthropological type the Sumerians of Mesopotamia and suggested that the mysterious Sumerians came from India. More recently Dr. Hüsing has drawn attention to a likeness between figures on early Buddhist carvings and those on Sumerian works of art. Whichever way the races drifted, an ethnic element common to India and Mesopotamia seems clear and to it might be ascribed the interrelated cultures.

Were it Sumerian, it could not be Aryan, but the simple equation is not yet established. The historical Sumerians did not use painted pottery, but seem rather to have displaced or conquered an older people who did; for instance, at Ur graves contemporary with the First (Sumerian) Dynasty have disturbed older interments accompanied by painted vases. But even if the culture common to the Indus and the Euphrates valleys belong to a "pre-Sumerian" stratum, it is still unlikely to be Aryan. Christian [2] distinguished in the Sumerians' monuments two racial types and in their language two components, neither of which is Aryan but one of which may well belong to the vase-painters. To this extent the attribution of the new finds to Aryans seems unlikely. The female figurines again do not seem proper to Indo-Europeans and the same types are found in South India as well as in the Punjab. On the other hand it should be recalled that a grave under a barrow near Belliah, Bengal, contained, besides apparently cremated bones and remains of a wooden pillar, female images impressed on gold leaf. The excavator would see in these the goddess Prithivī (Earth) to whom the Vedic Funeral Hymn (X, 18) commends the remains of the departed. [3] A final pronouncement must, of course, await the measurement of the new skeletal material and the decipherment of the script-signs on the seals and copper bars found in the Punjab.

The second possibility can only be judged when an examination of the new remains in their stratigraphical order determines whether a real break in culture is detectable when cremation first comes

[1] *Anc. Hist. of the Near East*, 1913, p. 173; cf. *Man*, xxv, 1.
[2] *M.A.G.W.*, liv.
[3] *Arch. Survey of India*, 1906–7, pp. 122 f.

in or at some other point. The last alternative might seem to be supported by the apparent discontinuity between the art, script, and other products of the prehistoric civilization and the creations of Aryan India. But there again the verdict must be suspended till further researches shall reveal whether the cleavage is absolute or whether the upper strata on the Indus sites may not serve to bridge the gulf. In any case it is in this area that the key to more than one of the riddles of human civilization lies hid and a bountiful reward awaits the excavations which alone can find it. Till then India offers but a tantalizing vista and its invasion by Aryans remains a fact to be inferred from linguistic data still disconnected from material remains.

5. *The Iranians in the First Millennium B.C.*

West of the Indians in the first millennium before our era dwelt the Iranians. Are the ascertained facts of their early history compatible with the account we have given of the Aryanization of India ? The oldest monuments of Iranian literature, the hymns or *gāthās* ascribed to Zoroaster (Zarathuštra), stand in a sense as a dividing line in space between the Indo-Iranians in Mitanni and the Vedic Indians in the Punjab. In the Avesta many of the Vedic and Mitannian deities have become devils while the name of Zoroaster's god, *Ahura*, is tending to mean 'demon' in the Veda. In this inversion we detect the hand of the prophet Zoroaster himself, who was perhaps the first great religious reformer. He took the old god, Varuna, who in the Veda is waning before the war-god Indra, raised him to a position of supremacy, stripped him of all material trappings and invested him with a sublime majesty as the guardian of the Cosmic Order (*Aša* or *Ṛta*). Some of the other popular gods of the Indo-Iranian period, such as Mithra, were retained in an attenuated form and subordinate position as personifications of abstract virtue. Others such as Naoṅhaitya (*Nāsatyā*) and (?) Indra were relegated to the armies of evil with whom the righteous man must fight on the side of Ahura Mazda but again as abstractions personified. Thus did Zoroaster convert the old Indo-Iranian polytheism into a spiritual monotheism which was no longer a mere tribal or national creed but a gospel to which all men were summoned.

The contrast between the Vedic-Mitannian religion on the one hand and the Avestan on the other is thus explained as the work of a dominant personality; it has even been suggested that the separation of Indians and Iranians was the result of a religious schism. In any case, spatial contiguity between Iranians and Indians seems implied, and that presumably after the Aryan conquest of Mitanni in the XVth century. Beyond this neither the age in which the prophet lived nor the scene of his labours are precisely determinable. His home is generally located in Sogdiana, Bactria or Arrachosia (Hara'uvatiš) and his reforms certainly antedate the accession of Darius I. Hall [1] and Jackson [2] would recognize in his patron, Vištaspa, Hystaspes the father of Darius. Eduard Meyer [3] on the other hand finds evidence in Median names such as Mazdaka, occurring in the Assyrian records as early as the VIIIth century, that Zoroastrianism was already established by that date. The prominence of the concept of Aša already in the XVth century B.C. as attested by the names of Mitannian and Syrian princes, would seem to favour the higher date. The people among whom the prophet worked must at all events be regarded as the eastern branch of the Iranian stock.

The West Iranian kinsmen of the Avestan Airyas begin to figure in historical documents about the VIIIth century. The first certain reference to the Medes [4] dates from that epoch when the Assyrians met them as far west as Lake Urmia. But the first regular monarchy under a line of kings with good Iranian names (Fravartiš, 'Uvakhšathriya, etc.) had its capital further east at Ecbatana. Then in the VIIth century the Persians under Teispes (Cišpiš) had established a dynasty among the Elamite Anshanites east of Susa to advance thence westward under Cyrus a hundred years later. In these peoples it is natural to see the western outposts of the Iranian population. Their appearance in history first in the region of Lake Urmia would be merely an accident resulting from the direction of Assyrian conquest and consequently of Assyrian geographical knowledge. And during the earlier part of the period covered by the Assyrian annals the more western parts of the highland north of Mesopotamia were certainly occupied

[1] *Ancient History of the Near East*, p. 555.

[2] *Persia Past and Present.*

[3] Article "Persia" in *Encyclopædia Britannica*, 11th ed. Darmsteter denies that Darius or any of the Achæmenids was a Zoroastrian. The name *Aššara Mazda* occurs in an inscription of Ashurbanipal (669–626 B.C.). *P.S.B.A.*, 1899, p. 132.

[4] The identification of the *Manda* mentioned by the Hittite kings about 1300 as living in western Armenia with the historic Medes seems questionable though it is accepted by Giles, *C.A.H.*, ii, p. 15.

by non-Aryan Asianic peoples akin to the bulk of the Mitannians of the Amarna age (page 19) while Iran proper does not yet figure in the cuneiform records. The monuments of these non-Aryan population are linguistically the "Vannic" inscriptions from Armenia, racially the Armenoid or Hittite types depicted on the bas-reliefs of Tell el-Halaf.

Nevertheless Hüsing, who localizes the Avestan *Airyanam vaêjaṅh* (Aryan homeland) in Armenia, contends that the Iranians entered Iran only during the first millennium B.C. while they had dwelt together with the Indians north of the Caucasus. The real answer to this contention seems to be supplied by the position of the earliest parts of the Avesta as contrasted with the Mitannian documents; the identity in difference is most readily intelligible on the assumption of a continuous population from Lake Urmia to the Punjab whose beliefs formed the material and background for Zarathuštra's reforms. This continuity should have been still subsisting at the time of the prophet and not broken off as Hüsing assumes somewhere in the middle of the second millennium when the Indians would have crossed the Caucasus. Secondly, the Mitannian and Syrian names in *Arta-* have, as we have seen, a distinctly Iranian tinge already in the XVth century.

Against these grounds for the belief in the presence of Iranians south of the Caucasus by 1000 B.C. have we any evidence for Iranians north of the range at an early date? In the VIIIth century a people called by the Assyrians *Azguzai*, the Scythians of the Greeks, crossed the Caucasus to descend upon Mesopotamia. Many people hold that these Scyths were Iranian. The linguistic evidence, limited to a few proper names mostly of late date, is inconclusive. Archaeologically, however, our people are well known. Undoubtedly Scythian art is strongly influenced by the Iranian —but it is not any specific Iranian art, Persian for instance, but has its own unique individuality. Again, the Iranian on a gold plaque from the Oxus Treasure (Plate II, 2) is wearing Scythian dress. The Scythian burial customs are, however, decisive. They are utterly different from those of the Iranians or Indians or any other Aryan people whatsoever. At the tomb of the chief his women and his servants were slain, and round the sepulchral chamber many horses were impaled. These rites, attested by the descriptions of Herodotus and by the actual remains from many a barrow, are utterly un-Aryan. On the other hand they find exact parallels among the non-Aryan Mongol nomads of Upper

PLATE II

1. DARIUS THE GREAT

2. IRANIAN IN SCYTHIAN DRESS
(British Museum)

[*face p. 38*

Asia throughout the ages as Minns [1] has amply demonstrated. This author treats the Scyths as Mongoloid forerunners of the Huns, Tartars, and Peschenegs, and that is no doubt right.

Professor Rostovtseff, who is the last to defend the Iranian hypothesis, has himself cut away the ground from under his feet. For he has convincingly distinguished the Sarmatians, who were demonstrably Iranian, from their predecessors the Scyths.[2] In their graves we find no more the horse hecatombs nor the heaps of slain women and retainers, but the simpler, albeit rich, funeral rites which would be appropriate to any other Aryan people. Tombs of the new type first appear in East Russia, in the Orenburg region in the Vth century, and spread gradually westward in the rear of the Scyths—to the Crimea in the IInd century and the Danube by 50 A.D. With this clear separation of Sarmatian from Scythian, the real ground for dubbing the latter Iranian—the Iranian names found in Scythia in Roman times and the language of the modern Ossetes—disappears.

At the same time Hüsing's attempt to treat the Scyths as European is misplaced. Scythian burials are found first in the east of Russia between the Kuban and the Dniepr. In the VIIIth–Vth centuries B.C. the material found west of the last-named river is quite different from the Scythian and is connected with the Central European Hallstatt-Lausitz series. It is only in the VIth century that the oriental rites and objects begin to intrude into West Russia and become established there two centuries later. There is indeed archaeological evidence for incursions of Scyths into Bulgaria, Hungary, and Eastern Germany, but only as invaders coming from the east.

But if the Scyths seem thus happily disposed of, there still remain the Cimmerians [3] to put in a claim for Iranian nationality. In Homer's time these people were presumably living somewhere to the north of the Black Sea and their name survived there in the Cimmerian Bosphorus into classical times. But the narrative of Herodotus combined with the data from Assyrian records shows that the Scyths drove some Cimmerians in a south-easterly direction across the Caucasus, where they appear as the *Gimiri*, north of Van (Urartu), while another band of them, mixed with Thracian Treres, fell upon Asia Minor from the West. Thus to split up the

[1] Minns, *Scythians and Greeks* (Cambridge, 1911), pp. 88 ff.
[2] *The Iranians and Greeks in South Russia* (Oxford, 1922), pp. 122 f.
[3] Minns, op. cit., p. 52; Rostovtseff, p. 40.

tribes of South Russia the Mongolian invaders must have been as far north as the Orenburg region where the Sarmatians first emerge, but need have been no further west. Now the claim of these South Russian Cimmerians to Iranian nationality rests upon the name of their chief, Sandakhšathra. This is certainly Iranian though the prince's father, Tugdammi, the Lygdamis of Strabo, has not even an Aryan appellation. But this one name appearing only after the Cimmerians had already been in Armenia for some time is but slender evidence for Iranians north of the Black Sea at an earlier date. I prefer Rostovtseff's view that the Cimmerians were an Aryan people indeed but akin to the Thracians.

Hence the evidence for Iranians in Europe before 600 B.C. has yet to be produced. The Scyths turn out to be Mongols, the Cimmerians Thracians. The first Iranians we can trace in our own continent are the Sarmatians, who come from the east, thus revealing how far the Iranian language and culture had spread already in the first half of the first millennium B.C. in Central Asia. And in Asia the Iranians appear in the Assyrian annals first on the north-eastern frontiers of the realm but under circumstances which allow us to infer a great hinterland of Iranians, not to the north but to the east. So the belief that this branch of the Aryan family was in occupation of the highlands of Iran before 1000 B.C. seems justified as a working hypothesis.

Archaeologically the earlier creations of the Iranian peoples cannot yet be disentangled from the general complex of Assyrian and Babylonian culture. Scientific exploration of early Median and Persian sites such as Ecbatana or Rhagae might have very important results. Already we can detect vestiges of connection between Iran and the West as early as the third millennium. They are represented by a beaked jug of Early Minoan type from Persia now in the Louvre. A rather similar jug is known from Anau in Turkestan a little further north and the same site has yielded spiral-headed pins and clay stamps which have their best parallels at Troy on the Hellespont. But we cannot yet say how much these phenomena are due to direct contact, how much to common reflections of the same Sumerian culture; a sickle from the same level at Anau to which a Trojan counterpart was long known has been shown by the recent excavations at Kish to be a Mesopotamian type.

In historical times the Medes and Persians rode on horseback;

the Vedic Indians normally harnessed their steeds to chariots and only exceptionally mounted on their backs. So the Iranians wore trousers, the Vedic Indians did not. These changes of fashion must then be posterior to the separation of the Indians and Iranians. As to burial rites, after the reforms of Zoroaster the dead were exposed on "Towers of Silence" as they are among the Parsis to this day; to avoid contamination of the sacred Earth or Fire inhumation and cremation were alike forbidden. But the Achaemenid kings were buried in rock-cut tombs and express prohibitions against cremation in the Avesta prove that that rite was also practised in Ancient Iran.

The racial type of the ancient Persians as represented by portraits of the Achaemenid kings is easily distinguished from that of Hittites, Semites, or Elamites by its lofty brow and delicate nose (Pl. II, 1). Only actual skulls can show whether we have here a Nordic or an Eurafrican type. The principal long-headed stocks in Iran to-day belong to the latter race. But the survival of blondes among the Kurds and around Persepolis to-day is a notable fact. And Chinese annalists speak of blue-eyed peoples in Eastern Turkestan about 100 B.C. The Iranians seem in fact to have reached the Tarim basin before the beginning of our era. Such an enormous range would suggest that they were partly nomadic. Yet in the *gāthās* of Zoroaster, although pastoral imagery abounds, the Aryan virtues are those of the husbandman, and the nomad is cursed as a Turanian robber. It must of course be remembered that the political empire of the Achaemenids and service in the Persian army resulted in the Aryanization of a multitude of heterogeneous peoples scattered far and wide throughout the vast domain.

We may now conclude that the Indo-Iranian peoples who appear on the north-eastern frontier of Mesopotamia with the Kassites about 1900 B.C. were but the advance guard of the great mass of the same stock. The western wing of these then reached Mitanni before 1500 B.C., while the eastern wing was descending into India not much later. The highlands between may be supposed to have been occupied by the people who a thousand years later enter the stage of history as Iranians, differentiating on the plateau from the original Indo-Iranian stock through admixture with non-Aryan Asianic and Turanian tribes. How the great mass of Indo-Iranian Aryans entered this region and whence they came cannot yet be decided; for the indications of direction viewed from this side are still too ambiguous. That question must await the inductive part of our inquiry.

CHAPTER III

THE ARYANIZATION OF THE MEDITERRANEAN

1. *Hellenes and Aegeans*

In the Mediterranean basin, the next area of the world's surface to be illumined by the light of history, there is again evidence of a non-Aryan population in lands that by classical times were Aryan. Place-names, a couple of inscriptions, a few traditions justify the belief in the presence of a pre-Indo-European stock on the southern coasts of Europe. But in this region history begins but late. On the other hand, the wealth of archaeological data is almost embarrassing.

In the Aegean the prehistoric past is illustrated by exceptionally rich and plentiful finds. But there is a gap in the archaeological record. We know on the one hand the historic Hellenic civilization beginning with the Early Iron Age or "Geometric" period. Behind this lies a dark age illuminated by little or no archaeological material, and on the further side of this chasm stands the prehistoric Mycenaean civilization which flourished between 1600 and 1200 B.C. and the many roots of which can be traced back to the fourth millennium on the one side in Crete, on the other through a series of distinct local cultures on the Mainland. Now the evidence of place-names abundantly demonstrates the presence of non-Aryan peoples on these shores.[1] For us the crucial question is: Did the Aryan element, let us call it the Hellenes, only intrude during the Dark Age or were there Hellenes also in Mycenaean and pre-Mycenaean Greece? I cannot pretend here to set forth in full nor to examine in detail the multifarious answers which have been suggested; space permits only of a most summary sketch.

If we interrogate Greek tradition, we find that the Hellenes preserved the memory of considerable movements in the population. Dryopes descended from Central Greece into the Argolid; Petthaloi advanced from the Epirus "to inhabit the Aeolian land"

[1] Kretschmer, *Einleitung in die Geschichte der grieschischen Sprache*; Fick, *Vorgriechische Ortsnamen*.

of Thessaly; Eleans swooped down from Aetolia into the western Peloponnese; above all the famous Dorians from the Pindus ranges or Macedonia occupied Boeotia, Laconia, the Argolid and Crete. But on the one hand these peoples are not represented as coming from outside the South Balkan peninsular; Macedonia is the furthest point to which tradition takes us. At the same time the Greeks regarded the peoples whom these migrants conquered as already Hellenic; the pre-Dorian population was not only Aryan, it is often called autochthonous. Save for a vague phrase about "the sons of Hellen being called to help the States" no reminiscence of a Hellenization but only one of a Dorianization survived in historical times. The other Hellenes might have been in their classical seats *ab origine*.

The evidence of language is partly concordant with the tradition in this form; it at least reveals two strata of Hellenic speech in Greece. The Hellenic dialects fall into four main divisions—Aeolic, Attic-Ionic, Cypro-Arcadian and West Greek. All these dialects are cast in the same mould. If not certainly sprung from a single Hellenic language, as Meillet [1] supposes, they are so intimately related that they must have been differentiated in strict contiguity, in a linguistic continuum. It seems for instance inconceivable to say Doric should have developed somewhere in the Danube valley while Ionic had been spoken in Attica a thousand years earlier. Be that as it may, the distribution of the dialects shows an intrusion of West Greek speech into an area previously occupied by other dialects. Arcadian has been left like an island in the Peloponnese cut off by a sea of West Greek dialects from its sister tongue in Cyprus. In West Thessaly West Greek elements obviously overlie an older Aeolic stratum. In Central Greece the West Greek dialects of Achaea, Phocis and Locris may have broken an older continuity between the Aeolic of Boeotia and East Thessaly, though the Aeolism of the former region is perhaps rather due to the invaders from Arne in Thessaly driven out by the West Greek Petthaloi.[2]

The most popular hypothesis is to connect this intrusion of West Greek speech into the east of the peninsula with the Dorian migration associating the movements of the Dryopes and Petthaloi

[1] "Tous les parlers grecs connus reposent sur une langue commune, déjà très diffèrent de l'indo-européen, à savoir le grec commun dont ils sont tous des transformations diverses."—*Aperçu d'une histoire de la langue grecque*, p. 18.

[2] Hdt., vii, 176, cf. Thuc., i, 12, 3.

closely therewith. But the reality may well be more complex. R. Meister proposes to recognize as the marks of Dorian speech not the general phenomena common to all West Greek dialects but certain peculiarities which were singled out for mention by the Attic comic poets.[1] If Meister's theory be accepted, it will follow that the West Greek invasion which isolated Arcadian was pre-Dorian; for our author regards even the non-Dorian perioeci of Laconia as West Greek. So we should have to reckon with a double migration of West Greek tribes. In any case, there are obvious difficulites in compressing into the Dark Age first the Hellenization of Cyprus and Arcadia, Attica, Boeotia and East Thessaly and then the conquest or isolation of these territories by a further invasion from West Greece.

Can we penetrate beyond the Dorian invasion? Behind it lies the Heroic Age, disclosed in the Homeric poems and later legends with an almost fabulous glamour and radiance about it. Yet recent researches have done much to establish Homer's credit as a source of historical information and to substantiate the golden age of tradition. The excavator's spade has exhumed the splendours of the epic citadels; T. W. Allen[2] has proved that the Homeric Catalogue of Ships describes a political grouping that could serve no ambitions of classical States, has no counterpart in "historic" times, but accurately reflects a real situation existing in a pre-Dorian epoch; and as the Egyptian and Hittite records begin to speak, the peoples and characters of heroic myth become historical.

But the interpretation of these sources and their co-ordination with archaeological results are still highly controversial matters. Homer describes a Greece in which the ruling dynasts in most States were "Achaean". The first question is: Were these Achaeans who led the expedition against Troy Hellenes? The Greeks themselves certainly thought so; Homer and later tradition depict them as worshipping the undoubtedly Aryan god, Zeus. Their outlook on life corresponds closely to that of an Indian

[1] "Dorer und Achäer," *Abhandl., d. phil.-hist. Klasse d.k. sächsischen Gesellsch. d. Wissen.*, xxiv, 3. His "Dorian shibboleths" are (i) the replacement of secondary intervocalic σ by h (Mωhα for Mωσα), (ii) the change of θ to σ (σῦμα for θῦμα), (iii) the assimilation of ζ to δδ, and (iv) the pronunciation of ε before ο- vowels as ι (σίω for θέω). All these peculiarities are detectable at Sparta and many of them also in Crete and the Argolid where Meister supposes the Dorians to have been but a small aristocracy. However, it is to be noted that (i) is also observed in Cyprus which is not usually classed as Dorian in any sense.

[2] *The Homeric Catalogue of Ships*, Oxford, 1921.

prince in the Veda or a Teutonic chief in the Norse epics. Yet many of their names—Odysseus, Achilles, Pelops—can only with the utmost difficulty and by torturing phonetics be explained as Indo-European. Later tradition brought the Pelopids, the Danaids and others with singular unanimity from Asia Minor and called them Phrygians, Lydians, or Lycians.[1] In other cases their affinities seem to lie in the South, in Minoan Crete, where a great civilization of seemingly Mediterranean type can be traced back uninterruptedly to the fourth millennium.

Sir Arthur Evans [2] long ago suggested the idea of a Minoan epic, written perhaps in the Minoan script and presumably in a non-Hellenic tongue. To explain the Minoan traits in the Greek epics he suggested that Homer might have incorporated episodes from the Minoan poem in his Hellenic rhapsodies. The recent discovery of scenes from heroic myth on Minoan-Mycenaean signet rings [3] may be interpreted in support of this view. And now Mr. Allen [4] comes forward with the theory of a Heroic Chronicle episodes of which were elaborated by his Chian Homer and another version of which survives in Dictys of Crete. It too might have been composed in the Minoan speech and written in the Minoan script. It is not inconceivable that the first Hellenic invaders should have appropriated pre-Hellenic national heroes and adopted as an ethnic title the tribal and local appellations given to the lands they occupied by earlier dynasts. It is equally possible that a pre-existing Hellenic population should have accepted and, in later tradition, Hellenized non-Hellenic culture heroes coming from Asia or Crete. Between 1600 and 1450 civilization was spreading from South to North and great royal tombs mark the establishment of kingly houses, whose ancestry would seem to lie in Minoan Crete, among more backward people on the Mainland.

To the solution of this question the Hittite records seem likely to give material help. If Tavagalavas, king of Aḫḫiyava about 1325 B.C., be really Eteokles (*'Ετεϝοκλεϝης*), there will be no doubt about the presence of Indo-European Hellenes in or near the Aegean area in the XIVth century; for this name is good Indo-European. And about a century later we read of an Alakšandu who seems to be an Alexander and so certainly Hellenic.

[1] See Weill in *Syria*, ii, pp. 135 ff.

[2] *The Minoan and Mycenæan Element in Hellenic Life*, *J.H.S.*, xxxii (1912), esp. p. 293.

[3] *J.H.S.*, xlv, pp. 27–41.

[4] *Homer: The Origins and Transmission*, Oxford, 1924.

But admitting that Homer's Achaioi are indeed Hellenes, there is a further question. Did they speak an East Greek dialect—Cypro-Arcadian, Ionic, or Aeolic or some more primitive tongue from which these three have developed—or was their language West Greek? In the former case they may well be the first Hellenes; in the latter it will be they who isolated Arcadian and an older Hellenic stratum will have to be assumed to account for Arcadian, Attic and Aeolic and some theory such as Meister's must be invoked to distinguish the later Dorians. The first view is of course the most economical. Moreover, Cypro-Arcadian elements are detectable in many regions where Homer or the Hittite records reveal Achaeans—Crete and Pamphylia for instance [1] —and where West Greek elements are lacking. Again, Cypro-Arcadian words in "Epic Greek" are noticeable, e.g., *κοίρανος*.

On the other hand Chadwick [2] has pointed out that the dialects of both the Achaeas were essentially West Greek and the inhabitants of these regions have a good claim to be regarded as descendants of the Homeric Achaioi since no tradition of the Dorianization of the areas survives. Moreover, parts of the Achaean realm—the Argolid, Messenia, Kos, Kalymna Rhodes—also spoke more or less West Greek dialects in historic times and cannot in all cases be regarded as Dorianized. The balance of probability seems to lie with Chadwick's view which has recently received the weighty support of Dr. Penrose Harland.[3]

Finally, were these Achaeans natives or intruders? As far as their dynasties are concerned, they are clearly newcomers; after four generations their lineage is lost; they "go up to a god", as Herodotus puts it. Moreover, they are described as having recently ascended the throne which they have in many cases won by marriage with the old king's daughter. Chadwick [4] has very aptly compared these and other phenomena of the Greek Heroic Age with the formation of Teutonic dynasties on the ruins of the Roman Empire. Contact with Roman civilization had broken down the bonds of barbaric society and permitted the emergence of personal war-chiefs endowed with a status and mentality very like that of

[1] E.g. *οἰ, ἰς, ἰν, πεδα, ἰος* in Crete. Cf. Thumb, *Handbuch der griech. Dialekte*, § 41. Bechtel, vol. ii, §§ 809–20, emphasizes the Cretan affinities of Pamphylian, but West Greek elements are also detectable.

[2] *The Heroic Age*, pp. 280 ff.

[3] *The Peloponnesos in the Bronze Age, Harvard Studies in Classical Philology*, xxxiv, 1923.

[4] Op. cit., pp. 353 ff.

Agamemnon or Achilles. At the same time service in the legions had educated the barbarians till they could usurp the domains of their former masters almost without a blow. A less probable explanation would be to regard the Achaeans as culture heroes, divine kings practising exogamy as Sir James Frazer [1] suggests, and therefore obliged ever to be seeking new kingdoms.

But though so obviously newcomers, the origin of the Achaean princes is obscure. Some as we saw come in the last resort from Asia; others, most notably Achilles' father, Peleus, the king of Phthia, come from the south, in this case Aegina; Diomedes of Argos had Aetolian antecedents; Atreus came to Mycenae immediately from Pisatis on the West coast; the Aeolid princes of Pylos in Triphylia hail from Thessaly.

New evidence has recently been announced from an unexpected quarter, but it rather complicates than elucidates the problem of the Achaeans. The Hittite king, Myrsilos, makes mention of a certain *Antaravas* who is described as king of *Aḫḫiyava* (Achaea) and *Ḷa.az.ba* (? Lesbos). Another king of Aḫḫiyava, perhaps the son of the foregoing, is named Tavagalavas, and bears the epithet *Ayavalaas* (? Aeolian, Αἰοϝολος). Dr. Forrer [2] identifies these two fourteenth century kings with Andreus and Eteokles, the reputed founders of the Minyan dynasty of Orchomenos. Nearly a century later one *Attaraššiyas*, king of Aḫḫiyava, plunders the coasts of Caria and Cyprus, and becomes an ally of the Hittite king. He and a helper receive the title *Kurivanies* (κοίρανοι). In this prince the German decipherer would recognize Atreus, Agamemnon's father. *Alakšandus* of *Uilusa* is mentioned by Muttallis a little earlier (1310–1290 B.C.).[3] The name of the Achaeans had of course been identified many years previously by Dr. H. R. Hall among the Sea Peoples who attacked Egypt under Merneptah about 1230 B.C., and more recently Autran [4] has suggested that it lies hid in the Biblical Hivites.

Now it is to be noted that all these peoples and the kings mentioned in the Hittite records are only certainly located in Asia Minor. From the standpoint of the lords of Boghaz-Keui the domain of the Achaeans seems to have lain in Pamphylia.

[1] *Lectures on the Early History of the Kingship*, p. 240.

[2] *MDOG.* lxiii (1924). Garstang and Mayer, *Index of Hittite Names, B.S.J. Suppl.*, identify Aḫḫiyava with Ἀγχιάλη.

[3] *Glotta*, xiii, p. 205.

[4] *Syria*, iii, p. 39.

Dr. Forrer, however, points out that their kings are important figures —peers of Pharaoh and the Assyrian and Babylonian monarchs—and concludes that Pamphylia was only the eastern corner of a kingdom the heart of which lay in Greece. Professor Sayce is dubious about this inference and sceptical as to the identifications of Andreus, Eteokles and Atreus, though he accepts "Achaea" and "Aeolian". It is therefore possible to argue that these Achaeans were an Asiatic people either preparing to conquer Greece or come thence expelled perhaps by the Dorians. Even on Dr. Forrer's own view the appearance of an Aeolid king of Achaea over a century before the Trojan War is distinctly puzzling. Still, despite such doubts and perplexities, these startling discoveries on the whole strengthen the belief that Hellenic dynasts were ruling in Greece by the thirteenth century and that they were pre-Dorian. Provisionally we shall adopt that position.

Such in their barest outlines and greatly simplified are the contradictory historical and literary data by which the archaeological material has to be interpreted. Let us begin with the Dorians as the most substantial figures of Greek tradition.

Despite the catastrophic effects of their descent, they have left singularly little unambiguous evidence of their inroad. Notwithstanding the Dark Age which intervened and the very obvious contrast between the Mycenean culture and the Geometric, a closer study of the remains reveals an even larger number of Mycenaean survivals in archaic and classical Greece. Moreover, the phenomena of the Iron Age in Hellas have many parallels in Asia (Cyprus, North Syria and Palestine), and must in part be explained from that quarter (see page 53 below). But tradition is sufficiently definite to justify us in looking to the north for objects to be associated with the Dorians. Now at Sparta, the centre of Dorian life in classical times, Mr. Casson [1] has called attention to certain objects which do point unambiguously in that direction. These are brooches or fibulae in the shape of double spirals, conventionally termed spectacle-fibulae (Fig. 8, 9) with which are associated curious horses and birds of bronze or clay. In the light of these brooches Mr. Casson can trace his Dorians along precisely the routes indicated by classical authors into Macedonia.

In the Vardar Valley and further west on Lake Ostrovo the same author has identified an Early Iron Age culture, the exact dating

[1] S. Casson in *Ant. J.*, i, pp. 200 ff.; *B.S.A.*, xxiv.

of which is not yet altogether clear,[1] associated with fibulae and bronzes of the types just described and painted "geometric" pottery. Mr. Casson is indeed too orthodox to see in these new-found strata the proto-Dorians; for them he looks further north —to the Hallstatt or proto-Hallstatt civilizations of Illyria and the Danube valley. And certainly the spectacle brooches may be derived from that area and some of the Macedonian pottery has Illyrian analogies as far as the vase-handles are concerned. Nevertheless an invasion from Central Europe does not seem a necessary postulate though it remains a possibility to be kept in mind.

Neither in Macedonia nor further south do we find anything like a bodily transplantation of the Illyrian or Danubian cultures

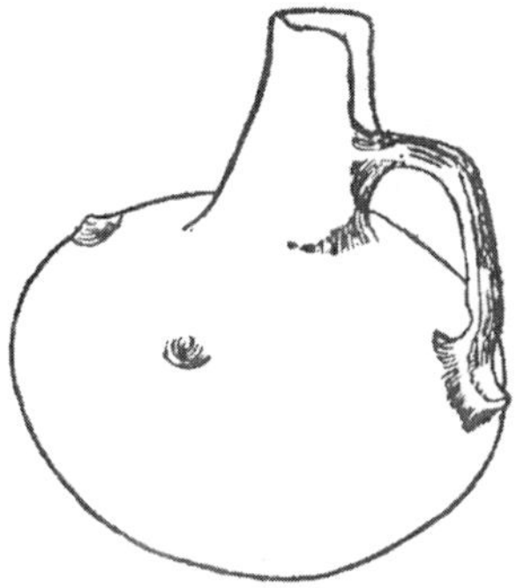

FIG. 4. Early Iron Age Jug.

of the Hallstatt or immediately preceding epochs. The Iron Age pottery from the Vardar valley is not Hallstatt or any other Central European pottery; it is on the contrary deeply rooted in a Balkanic tradition, which once was indeed observed in a wide circle through the Danube valley, Illyria, and even Upper Italy, but was already assuming a local character in the south-west Balkans by 2000 B.C. The forms of the vases, jugs with cut-away necks (cf. Fig. 4) and goblets with high handles, and even the fantastic handle-types which suggest Illyria can trace their lineage in Thessaly back to the end of the third millennium (in the third "pre-Mycenaean" period contemporary with Early Helladic, below, page 59). The painting, which is quite distinct from that of the Hallstatt school,

[1] It begins to appear on the Vardar immediately after the fall of a settlement containing the latest style of Mycenaean ware current in the XIII–XIIth centuries. Heurtley, *L.A.A.*, xii, p. 35.

is intimately related to that of "sub-Mycenaean" wares in Thessaly (Fig. 6, 5) which it would seem also reached the Vardar itself. A cultural contact would suffice to explain the bronzes.

I would therefore be inclined to regard the Early Iron Age culture of Macedonia as a native development of a much older South Balkan stock which had borrowed certain elements from its Central European neighbours. The Dorians were among the authors of this civilization and carried its traditions with them to the South, driven out perhaps by the pressure of Illyrian and Celtic tribes to the north and west.

The Achaean problem is yet more intricate. Its solution still depends upon our attitude to the Homeric poems. On the one hand the political geography and the civilization of the Achaean period as depicted in Homer correspond most closely with those of the Mycenaean age. The homes of the heroes are sites which possessed in the Mycenaean period an importance they never subsequently enjoyed. The glories of Mycenae and Nestor's Pylos described by Homer were realities in the sixteenth and fifteenth centuries. And Mycenaean civilization spread northward just as some Achaean princes did. Not only so, at Mycenae we may distinguish two dynasties—an older line of kings who lay buried in the celebrated Shaft Graves, and a later house whose scions built stately beehive tombs and whose accession coincides with the greatest expansion and wealth of the Mycenaean world.[1] Just so tradition tells of two dynasties at Mycenae—Perseids and Pelopids! Thus the legendary figures of the Heroic Age seem to become flesh and blood as culture-heroes who civilized a barbarous Hellas. Can we wonder that T. W. Allen wrote: "The 'nameless' Mycenaeans were the Achaeans"?[2]

On the other hand, the contrasts between the "Achaean" age and the Mycenaean are notorious and are growing more, not less, glaring. The expansion of the Mycenaean civilization reached its culmination by 1400 B.C.; the earliest mention of Achaeans dates from the end of the fourteenth century, and, if Forrer be right, Atreus reigned a century later still. Nor does the distribution of Mycenaean remains and the centres of Mycenaean life coincide so perfectly as has been suggested with the Achaean sites in the Homeric Catalogue. Let me take but one instance, Allen's crucial

[1] See the account of the latest excavations by Mr. Wace in *B.S.A.*, xxv, p. 120. Sir Arthur Evans, however, maintains that the beehives like the Shaft Graves go back to period M.M.III. *J.H.S.*, xlv, pp. 45 and 75.

[2] *C.R.*, xxv, p. 234.

example of the spread of culture northward. In the kingdom of the Aeginetan Peleus in the Spercheios valley, where alone "Achaeans" and "Hellenes" are used as tribal appellations by Homer, not a single Mycenaean tomb has been found.

Finally, however closely the civilization pictured by Homer corresponds to the Mycenaean, the familiar discrepancies remain. The Mycenaeans normally used huge shields covering the whole body, shaped some "like a tower", some like a figure 8, but no

FIG. 5. Sherd of Achaean (Late Mycenaean B) ware from Tiryns.

body armour (Pl. III, 2); some of Homer's Achaeans, instead of these large shields, carried a round targe and wore breastplates. The true Mycenaean swords were all designed exclusively for thrusting (Fig. 25, 1 and 2); Homer describes a certain number of undeniable sword strokes implying a slashing weapon. In the Mycenaean age iron was only used for ornaments; the Homeric age was a bronze age, too, but passages in the poems mention iron tools and cannot be explained away. And lastly the Mycenaeans were always interred in corbelled vaults or rock-cut chamber tombs; the Achaeans in Homer practised cremation.

Hence some authorities, Sir William Ridgeway,[1] Dr. Mackenzie,[2] and Professor Chadwick,[3] have sought to find, after the pure. Mycenaean age of the sixteenth–fifteenth centuries, a transitional period which while preserving the essential outlines of the Mycenaean shall yet exhibit those innovations which distinguish the Achaean world. And as a matter of fact we do find in very late Mycenaean deposits illustrations of breastplates and round shields (Fig. 5 and Plate III, 1), cutting swords (Fig. 25, 3–4), and occasional cremations. With these are often associated foreign objects such as brooches or fibulae of very simple type (Fig. 8, 1–4). At the same time a change is observable in the style of the ceramic decoration from the free style with naturalistic motives taken from marine or plant life of the Mycenaeans to a metopic arrangement which in its striving after symmetry and balance seems to foreshadow the distinctive features of classical vase-painting and in which the introduction of human figures betokens a new interest in man.[4] Moreover, the forms of the metope style vases are in some cases strange to the pure Mycenaean repertoire. And a recently found "treasure" of this epoch contained a Hittite cylinder [5] suggesting just that sort of contact by raids which the Hittite records attest (see note at end of chapter).

It must be admitted that this period, which we shall provisionally term "Achaean", is still vague and that its aspect still only partially coincides with the picture given by Homer. To the wealth of the epic kings we find in it no parallel; it is a period which gives every sign of exhaustion, poverty and decline. The burial rites are far from Homeric; inhumation was still the rule, and in the rare cases of cremation the ashes were laid in urns in the old-fashioned chamber tombs. Only at the very end of the period do we find at Halos in Thessaly [6] pyres surmounted by barrows which do correspond exactly to the rites described by Homer, but here the weapons were of iron, which is un-Homeric. Still the assumption of an Achaean period as thus defined seems the most hopeful way of escape from our dilemma.

[1] *The Early Age of Greece.*
[2] *B.S.A.*, xiii.
[3] *The Heroic Age*, pp. 185 ff.
[4] Dr. Penrose Harland has failed to notice the very real change in the pottery that characterizes the latest Mycenaean epoch and so falls into the error of assigning fibulæ to the Mycenaean period without qualification. All the examples found in a definite context are "Achaean".
[5] *'Αρχ. Δελτ.*, 1916, *παραρτήμα*, pp. 13 ff.
[6] *B.S.A.*, xviii.

PLATE III

ACHAEAN AND MINOAN ARMAMENT

[*face p. 52*

Whence then came the innovations? To this question no single answer can be given. It is in fact certain that the "Achaean" period begins in Greece with no sudden break in culture; no catastrophe disturbed the old sites. Still less was the Mycenaean civilization superseded bodily by another; the change is quite gradual, the new elements came from different quarters. The industrial use of iron is now proved by the cuneiform documents to have begun in Asia Minor and presumably our Achaeans got their iron from that quarter. In this respect Homer was a good archaeologist; for he makes iron common only among the Trojans or among the Achaeans after they had been plundering Asiatic towns. It is possible that the rite of cremation came from Asia too. It was indeed the practice in Central Europe by the Middle Bronze Age (from about 1450 B.C.). But the earliest Aegean cases come from Caria or insular regions in close touch with Asia—Crete and Salamis; the very late Mycenaean tombs with Achaean ware and fibulae in Cephallenia, Achaea (Patras) and Boeotia (Thebes) contain, so far as we know, only unburnt bodies. In later days too the rite was most regularly practised within the sphere of Asiatic influence, e.g. in Thera and Crete; early Geometric cemeteries in the Peloponnese at Argos, Asine and Tiryns show no cremations. On Aegina and at the Dipylon near Athens cremation was rarer than inhumation. In any case, as we shall see in Chapter VI, Ridgeway and Rohde tend to exaggerate the significance of the rite.

Finally the new tendency in ceramic decoration—the division of the surface into panels or the metope style—has very ancient precursors in Hither Asia (cf. Fig. 15) and was most richly developed on the Philistine pottery of Palestine (Fig. 6, 3–4) and on contemporary Syrian wares.

Nevertheless the Oriental influence should not be over-estimated nor the scope of the Asiatic parallels exaggerated. The iron and the Hittite cylinder betray contact between Greece and Anatolia just as do the cuneiform records. That does not mean an Oriental invasion. The metope pottery in Palestine is usually regarded as a foreign fabric introduced from the Aegean by the Philistines,[1] and if M. Autran's suggestion as to the Hivites be correct, they too may have been colonists who assisted in the propagation of the

[1] See Pythian-Adams in *B.S.J.*, iii. Saussey contests this view, arguing that the "Philistine" painting and motives are rooted in an ancient Asiatic school exemplified, e.g. at Susa; *Syria*, v, p. 184.

strange ware. In Greece itself the metope style may go back to the pre-Mycenaean period. It is in any case certain that the innovating vase-forms that characterize the Achaean epoch have their history in pre-Mycenaean Hellas; the most notable Achaean shape, a bell-shaped crater (Pl. III, 1 and Fig. 6), may be traced back

FIG. 6. Bell-shaped craters: 1–2, Greece (Achaean); 3–4, Palestine (Philistine); 5, Halos (Proto-Geometric); 6, Asine (Geometric) (after Pythian-Adams).

on the Mainland to that same South Balkan culture, the persistence of which we have detected in Iron Age Macedonia. The same is still more obviously true of the jugs with cut-away necks that appear in pre-Dorian Thessaly as they do in proto-Dorian Macedonia.

On the other hand, two of the phenomena of the Achaean period point unmistakably to the north or north-west. These are the fibulae and the slashing swords which were undeniably invented in the Danube valley or further north.[1] I do not, however, think that the intrusive weapons necessarily betoken a wave of conquerors coming from Bosnia or Hungary as Ridgeway and Peake [2] imagine. That is indeed a possibility, but we should then expect to find many more objects of Central European type in Hellas; by 1350 B.C. Danubian civilization[1] was characterized by very distinctive pottery and splendid bronzes. Achaean pottery is no more Danubian than proto-Dorian pottery was Hallstatt and the continental bronze types, apart from the fibulae and slashing swords, are even more conspicuously absent from the Aegean. I repeat, the Achaean period does not reveal the older culture as abruptly superseded by any other. As in the case of the Macedonian Iron Age a culture contact with the north and a tribal movement within the Balkans will account for the innovations of the period.

At the same time we have seen that the phenomena of the Heroic Age presuppose contact between its authors, the relatively barbaric Achaeans, and the higher civilization of Mycenae. And that contact most probably took place in the Balkan peninsula itself on the fringe of the Mycenaean civilization. We might perhaps recognize in a serving man who is painted white among the red " Mediterranean " Mycenaeans on a Tiryns frescoe,[3] a precursor of the Achaean conquerors in just that position occupied by the Teutons during the epoch of their education by Rome. The habitat of the Achaeans in their period of tutellage would on *a priori* grounds be located in north-west Greece—Epirus, Aetolia, Acarnania and Levkas. The dialect evidence on Chadwick's view points that way, and it is on the Adriatic coasts, traversed since 1600 B.C. by the ships bringing amber from the north, that the use of slashing swords and brooches could most easily be learned.

This area is still inadequately explored. It does not seem to have been Mycenaeanized, but it looks rather as if a backward culture akin to that of the Iron Age in Macedonia and rooted in the same pre-Mycenaean culture which appeared in Thessaly shortly before 2000 B.C., had embraced all north-west Greece, while stray Mycenaean imports there, e.g., a sword from Dodona, illustrate the requisite contact with the higher civilization

[1] Childe, *Dawn of European Civilization*, pp. 198 and 216.
[2] *The Bronze Age and the Celtic World*, p. 112.
[3] *Tiryns*, ii, p. 118 and pl. xi, 6. This interpretation was first suggested by Sir Arthur Evans.

of the Peloponnese. It is interesting to note that a curious culture with geometric painted pottery did in fact intrude into the Spercheios valley, the home of Hellenes and Achaioi in Homer, coming apparently from the west across Tymphrestos. We may then regard the Achaeans of the XVth century and earlier as among the authors of the West Greek cultures vaguely known at Levkas and Lianokladhi on the Spercheios and others as yet undiscovered. Then they will be a southern wing of a long series of West Greek tribes of which our proto-Dorians in Macedonia will represent the northern flank. The close kinship between Doric and the West Greek Achaean dialects will then be explained.

If this view be correct, the Achaeans were not the first Hellenes in Hellas; for they turn out to be West Greek. There must have been Hellenes in the Peloponnese already before the Achaean dynasts usurped the thrones of Mycenae, Tiryns and Lacedaemon. We have then still to find the Arcadians, Ionians and Aeolians.

Our Achaeans were the heirs of the Mycenaean civilization. The latter occupied the whole of the Greek Mainland except the Spercheios valley, inner Thessaly and north-west Greece during the XVth century, overlying older local cultures. Materially the Mycenaean civilization proper is just the Minoan civilization of Crete transplanted. Minoan art, religion and writing [1] were imposed upon the native "Helladic" cultures in such a way that an actual conquest and colonization by Cretans seems implied. Survivals of an indigenous culture are indeed everywhere observable during the Mycenaean age. There is moreover a residuum of phenomena in the period which cannot as yet be explained from Crete (the beards and sculptured stelae of the Shaft-Grave epoch, the beehive tombs of the next phase and the so-called megaron house of L.H. III). These native survivals and unexplained peculiarities are, however, insignificant in comparison with the Minoan elements. The religious symbolism from the tombs and palaces reproduces to the smallest detail the ancient Cretan cults. The art of the pottery, gems and gold-work is purely Minoan. The frescoes of the Mainland palaces must have been painted by artists from the island.

[1] Dr. Harland entirely fails to appreciate the significance of these phenomena and has altogether missed the Minoan inscriptions; only a few of the latter have been published but there are plentiful examples in the museums of Navplia and Thebes: see also *J.H.S.*, xliv, p. 275; *Scripta Minoa*, i, p. 57. On the pottery see Forsdyke, *B.M. Catalogue of Vases*, i, p. xxxix.

They depict women in Minoan costume, and the men are the red Mediterraneans familiar from the walls of Knossos. The ancient Minoan script, or a dialectic variety thereof, was used for inscriptions on locally manufactured vases at Tiryns, Mycenae, Thebes and Orchomenos. And these manifestations of a new inspiration appear in the palaces and great tombs which evidently belong to new dynasties.

All this is best understood as the result of the establishment of Minoan princes on the Mainland and a real colonization of Hellas by Cretans who need not, however, have been very numerous. The question therefore arises were these Minoan colonists and dynasts, *quā* Minoan, Indo-European ? The Minoan civilization

FIG. 7. Introduction of the horse into Crete.

in Crete evolved continuously from the IVth millennium, and was in its essence Mediterranean. From first to last it looked un-Aryan. Only the decipherment of the Minoan texts can really decide that point, but, if we may judge by the adaptation of it in the Cypriote syllabary, the script had not been devised to express an Indo-European language. Minoan religion again with its predominant Mother Goddess does not recall the Aryan pantheon modelled on a patriarchal earthly society. The survival in Crete of languages and cults strange to the rest of Greece supports the same view. Finally the horse, which we shall see reason to regard as a specifically Aryan animal, was only a late importation into the island (Fig. 7).

Of course, Crete was subject to periodical intrusions from various quarters. During the period known as Early Minoan, 3300–2200 B.C., a brachycephalic element, possibly of Anatolian extraction, steadily increased in numbers and some of the later Minoan princes belong to the Anatolian type; at the close of the Early Minoan age there is evidence of a strong current of influence from more northerly islands in the Aegean. Finally at the end of Middle Minoan II the Cretan palaces were destroyed, perhaps sacked, only to rise again in the succeeding epoch. At any of these points a new dynastic element and with it a new language might have been introduced. Nevertheless, the development of culture was essentially continuous till Achaean ware came in about 1250 B.C. The non-Aryan characters affect Minoan culture as a whole. Hence with all due reserve we do not regard it as the work of Aryans. Consequently the Minoan princes and colonists who established themselves in Hellas between 1600 and 1400 B.C. cannot have been Aryans by birth. They may be responsible for those non-Hellenic place-names which Fick terms Hattid and Eteo-Cretan. But being few in numbers the invaders may in the end have adopted the speech of the pre-existing population along with the continental type of house.

The same sort of argument as applies to Crete allows us to eliminate from among the claimants to Hellenism the ancient population that created the Cycladic culture on the Aegean islands. And classical writers knew that those islands had been inhabited by non-Hellenic barbarians such as Lelegians and Carians. This verdict also disposes of one pre-Mycenaean stratum of people on the Greek Mainland. The maritime Early Helladic people who occupied the Peloponnese, Central Greece and Levkas between 2500 and 1900 B.C. were virtually identical with the islanders. To these maritime intruders may be ascribed the pre-Hellenic topographical names of the Mainland and the Western isles which Fick calls Lelegian on account of their Cycladic-Anatolian parallels. So neither the Mycenaeans of the XVI–XVth centuries nor the Helladic folk of the IIIrd millennium are likely to have been the Hellenizers of Greece.

But the Mycenaeans found the Helladic colonists already overlaid by a different racial and cultural stratum and the former had themselves to conquer older inhabitants of Hellas. Can either of these layers be identified as Aryan?

The archaeological record on the Greek Mainland begins in the IIIrd, or at the end of the IVth, millennium with a population of peasants living in Thessaly, Central Greece and Arcadia whose most conspicuous industrial achievement was a magificent painted pottery. Nothing in Greece itself proves that the neolithic population was continuous with the Hellenes, but in a later chapter we shall have occasion to ask whether these and other peoples who painted their vases were Aryans. In parts of Greece in any case these peasants were succeeded by a new band of vase-painters coming from beyond the Balkans who introduced the spiral motive and a new type of house, the megaron, which henceforth survived in Greece to become the plan of the classical temple *in antis*. It is undeniable that this new intrusive folk may represent the Hellenic element in the population of classical Greece. It is not, however, certain that their culture is effectively continuous with that which we have assigned to the Dorians and Achaeans. Their wider affinities will occupy us more closely in Chapter V. Here we must note that Fick recognizes a Thracian element in Greek topographical nomenclature. Now the second band of vase-painters came to Thessaly from Thrace and had relatives in Bulgaria. It is therefore tempting to attribute the Thracian names to them (we shall see that "Thracian" does not necessarily mean Aryan).

About 2300 B.C., or soon after the Early Helladic invaders had reached the Peloponnese, a third change is noticeable in Thessaly. Painted pottery went out of fashion and new types of vessel—high-handled cups and jugs with cut-away necks—came into use, and a tendency to fantastic elaboration of the handles made itself felt. At the same time the first perforated stone axe-heads and mace-heads made their appearance. This culture in a general way forms part of a huge province extending right across the Balkan peninsula from the Dardanelles to the Adriatic with ramifications in the Danube valley, Upper Italy and even Apulia. On the other hand the ceramic forms show this culture in Thessaly to be continuous with the proto-Dorian culture of Macedonia, and the Achaean material in Levkas, while locally it survived throughout the Mycenaean age to form the basis for the Early Iron Age of Thessaly itself in the Achaean period. It was in fact the substratum from which all three developed.

It had extended its sway at some time not yet precisely determined to West Greece where its types seem to succeed the Early Helladic forms in Levkas while cognate shapes are known from Cephallenia and the acropolis of Nestor's Pylos. Not only so, the third Thessalian culture was one of the parents of the so-called "Minyan" or Middle Helladic culture of Central Greece. The authors of the latter ousted the Early Helladic settlers from Orchomenos about 1900 B.C. and dominated the whole region till the advent of Minoans in the XVIth century. The same people laid violent hands on part at least of the Peloponnese—Korakou and other villages near Corinth, Argos, Mycenae, Tiryns—and on Attica and Aegina. Without, however, annihilating the Early Helladic culture in the south, these "Minyans" were the ruling caste till the first Minoan conquerors seized Mycenae about 1600 B.C., and continued to play an important part in remoter sites like Korakou even into the Achaean period. Thus soon after 2000 B.C. the Balkan peninsula possessed a degree of cultural unity not hitherto attained and not repeated after 1500 B.C. This cultural unity may reflect the linguistic continuum from which the Hellenic dialects of historic times were differentiated.

It is in any case certain that many of the ceramic forms that distinguish the later cultures which we have already indentified as Hellenic can be traced back to this epoch of uniformity; for instance, the bell-shaped crater which was so characteristic of the Achaean period is a common form in the "Minyan" ware of Central Greece and recurs in an allied fabric in Aetolia. The proto-Dorian jugs with cut-away necks and the fantastic handles of Macedonia have already been noted. Thus the Achaean and proto-Dorian cultures belong to the same people as made the culture of the third period in Thessaly. The continuations of the same culture throughout the Mycenaean period and the products of "Minyan" stragglers in Thessaly itself may then represent the activities of ancestors of the Aeolians. Finally the southward extensions of "Minyan" civilization to Attica and the Peloponnese will have Hellenized the Early Helladic folk and produced the ancestors of Ionians and Arcadians.[1] The Minoan conquerors did not, I assume, destroy the language of their subjects and may have ended by adopting it. In any case, it seems as if the colonists of Cyprus, who must have been Mycenaean rather than Minoan in the strict sense, took with them the Cypro-Arcadian dialect and that at a time when the Minoan script was still in use.

[1] My conclusions are here in harmony with those of Dr. Harland. But I do not suppose that the "Minyans" entered the Peloponnese speaking Arcadian as such; the differentiation of the three "East Greek" dialect groups would have taken place between 1800 and 1400 B.C. as a result of admixture with various non-Hellenic elements. Nor can I admit that the pre-Achaean Hellenes in the Peloponnese worshipped Poseidon to the exclusion of Zeus, since the latter was an Aryan god. Many hold that Poseidon was pre-Hellenic—Minoan or at least Aegean.

On the view here advanced then the first demonstrably Hellenic people were those who created the third culture of Thessaly.[1] In that case the proto-Hellenes as such emerge about 2300 B.C.

2. *The Thracians and the Phrygians*

North of the Hellenes dwelt from Homer's days various tribes somewhat loosely described by the classical authors as Thracian. Of their language we possess but a few late glosses and proper names. This scanty material is held to prove the presence of an Aryan element in the population. From the curious social customs and religious practices reported by classical writers it may be legitimate to infer that the Indo-Europeans in Thrace were much mixed with extraneous elements.

The culture of Thrace in the chalcolithic epoch, as it is called, is well known. Its basis was the same as the second culture with painted pottery that intruded into Thessaly about 2600 B.C. But it was very much mixed with other elements, some derived from the Danube valley, others such as phallos worship from Anatolia, others again such as stone battle-axes from the north and east. This mixed civilization of barbaric peasants persisted in those secluded valleys for many centuries. It may even have lingered on into classical times. As its authors, like the Thracians of history, painted their persons, they may well have been in some sense Thracians themselves. That they were yet Aryan does not, however, follow automatically. Still the chalcolithic civilization[2] is all that is known in Thrace till the end of the IInd millennium B.C.; a true Bronze Age is as yet undiscoverable.

The first evidence of a distinctly intrusive culture belongs to the full Iron Age. Then the presence of newcomers is denoted by graves containing spectacle-brooches and other objects such as the "Glasinac" fibulae of Fig. 8, 7, more or less reminiscent of the Central European Hallstatt civilization.[3] Stray bronzes of the same general affinity, such as socketed celts and "antennae" swords

[1] This view is provisional only. I am conscious of difficulties which I have not raised here. The possibility that all the Hellenes came in during the Dark Age still remains. Reservations have also to be made in respect of the claims of Asia; Macedonia, and western Asia Minor, which are key-areas, have only been scratched by scientific exploration and may yield unexpected results. Even more significant should be a proper study of the virgin soil of Epirus and Albania. Pending these researches I offer the above in all due humility as the most consistent synthesis of literary and historical data possible.

[2] The chalcolithic material is described in Childe, *Dawn*, ch. xi.

[3] *Izv. Bulgar. Arch. Instit.*, i, pp. 32 ff.

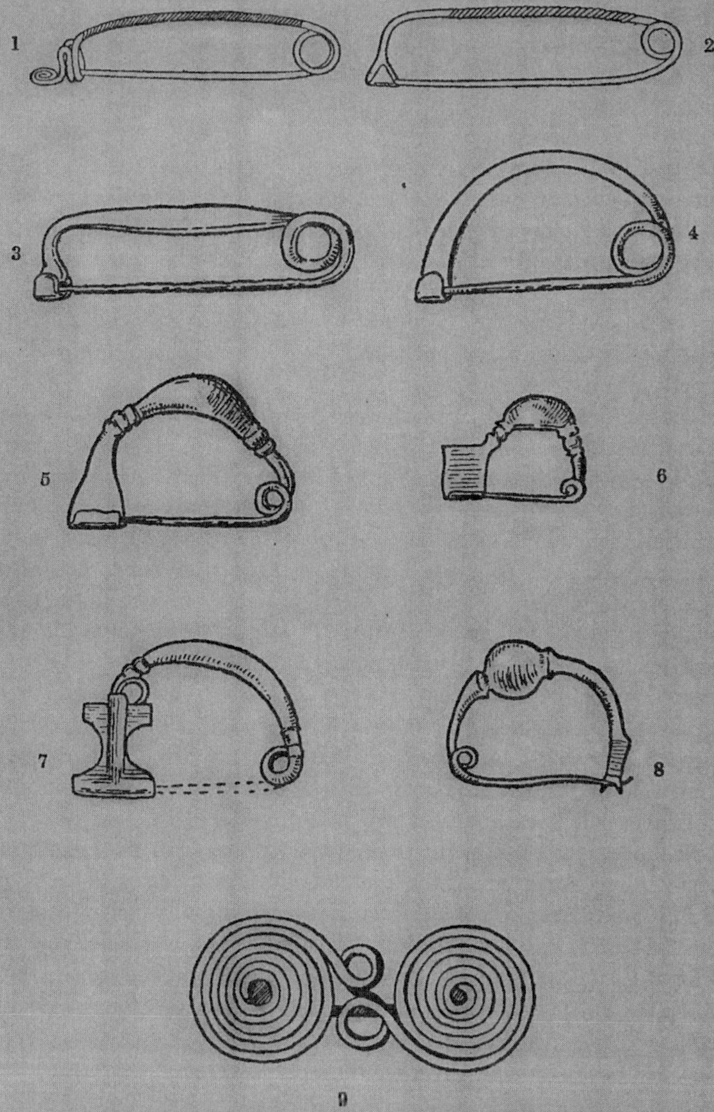

FIG. 8. Brooches: 1–2, Italy (XVth–XIIIth century); 3–5, Crete (Vrokastro chamber-tombs); 6, Crete (Vrokastro bone-enclosures); 7–8, Bulgaria; 9, Macedonia. For the distribution in Greece see p. 77.

(like Fig. 25, 6) found in Bulgaria, would allow of a much stronger case being made out for an invasion of Thrace from Central Europe than is possible in the case of Greece. But, even so, we look in vain for the Thracian swords which were renowned in Homer's time. To decide whether the Aryan element in Thrace is represented in the chalcolithic culture or one of its constituent elements or whether it only came in with the Iron Age must await later chapters. At present those are the possible alternatives.

Across the Straits in the north-west corner of Asia Minor we find, besides the coastal Greeks, a group of Aryan tribes who seem to stand out against an apparently Asianic background and who are connected by tradition and the evidence of names with Thrace. The most important of these were the Phrygians, who under Midas ruled an extensive empire in the VIIIth century B.C. Enough remains of the Phrygian language [1] to show that it was an Indo-European tongue. It exhibits in some respects close affinities to Greek and even made use of Greek words, but at the same time Slavonic parallels have been noted and certain features would connect Phrygian with Hittite Našili. The Phrygians, moreover, worshipped Aryan deities: their chief god, *Bagaios*, as well as the Moon-god, *Mēn*, has a good Indo-European name. On the other hand the great prominence of the mother goddess in their pantheon and references to matriarchy among their social institutions are quite un-Aryan features. Dr. Hall's conclusion [2] seems to be well-founded: "We may conceive of the Phrygians as a people compounded of an Aryan aristocracy ruling over and gradually mixing with the Anatolian peasants."

Now tradition consistently maintained that the Phrygians had come from Thrace, and there was in fact a tribe in the latter region whose name *Briges* seems just a deaspirated form of *Phryges* (**Bhruges*). But this migration must have taken place before the end of the IInd millennium, since in the Homeric Catalogue the Phrygians are mentioned among Priam's allies in his war with the Achaeans (about 1200 B.C.) and in such a way that they must have come from the classical Phrygia. At the same time Homer's account of the ethnology of the Troad is confirmed in a satisfactory way by a list of the Hittite allies encountered by Rameses II in his Syrian campaign of 1287 B.C.[3] Pharaoh boasts of having

[1] Hirt, p. 598; Ramsay in *J.R.A.S.*, 1883.
[2] Hall, *Anc. History of the Near East*, p. 476.
[3] See Pythian-Adams, in *B.S.J.*, i; some readings are very doubtful.

defeated the ? *Iliunna* (or *Ariunna*), *Derden, Luka, Pedes, Kelekesh, Mesa,* and ? *Mawunna.* These contingents in the Hittite army correspond very well to the Troes of *Ilium,* the *Dardanoi,* the *Lukoi,* the Leleges of *Pedasos,* the *Kilikes* of Thebes, the *Musoi* and the *Maiones* mentioned in the Iliad. Hence the Egyptian evidence provides one more proof of the value of Homeric data for the political groupings existing in the IInd millennium. We may therefore admit that there were Phrygians in Asia Minor about 1200 B.C.

But the Homeric geography of the Troad raises other problems. In the first place the relative compactness and solidarity of the population of this corner of Asia Minor over against the rest of Anatolia deserves note. It is in harmony with the silence of the Hittite archives and the absence of Hittite monuments which show that this region had escaped the domination of the lords of Boghaz-Keui. Secondly the discrepancies between the heroic and the classical geographies of the Troad suggest a displacement of peoples southward; Strabo [1] preserves some memory thereof. The causes of the dislocation are to be found not only in the convulsions consequent upon the victory of the Achaean assailants, but also in the inroad of the Treres from Thrace (p. 39 above). Finally the question arises: were other members of Priam's confederacy besides the Phrygians, Aryan or led by Aryans? In the case of some of the tribes mentioned, Leleges, Pelasgians, Cilicians, a negative answer seems inevitable. On the other hand, Homer gives heroes on the Trojan side good Hellenic names, more Greek in fact than those borne by many Achaeans. It may, of course, be that the Greek poet gave these personages Greek names much as Shakespeare gives some of his characters English names, Quince, Dogberry and so on, though they be Athenians or Sicilians. However, it is curious that the names in question belong very largely to a particular group of Hellenic appellations, namely those current principally in North Greece, Thessaly, Epirus, and above all Macedonia.[2] It really looks as if, besides the Asianic stocks, such as Leleges, and the Aryan Phrygians, there was another Aryan element in the ruling classes of north-west Asia Minor, and as if it was Hellenic or closely akin to the Hellenes. That is by no means inconsistent with the traditions connecting Mysians, Dardanians and Bithynians with North Balkan peoples when we recall the northerly

[1] See Leaf's edition, *Strabo on the Troad*, pp. 250 f. and 303 f.
[2] *J.H.S.*, xxxix, pp. 62 ff.

extension and Troadic connections of the culture which we call proto-Hellenic.

What then has archaeology to say on this topic and especially on the western connections? In the architectural monuments of classical Phrygia certain Mycenaean reminiscences may be detected: the heraldically opposed lions recall the Lion Gate of Mycenae. But these might be survivals of a very ancient tradition common to both sides of the Aegean. In the Phrygian barrows of the VIIIth or VIIth centuries excavated near Gordion,[1] the capital of Midas, fibulae or brooches were the most westerly objects discovered. But these are not derived from the Early Iron Age types worn in Thrace, but represent developments of the older pattern with a simple bow like Fig. 8, 4, which appeared in Hellas during the sub-Mycenaean period and is also known from the Caucasus, but is strange to Thrace. On the other hand, among the vases from the Phrygian barrows occur types such as jugs with cut-away necks, which we have already met in Thessaly and Macedonia. These again may be but continuations of much older local patterns, as is certainly the case with another curious vessel found with them—a beaked jug with a strainer in the neck used for decanting the national beverage, beer. The pedigree of this class of vase certainly goes back locally to early in the IInd millennium. Thus the most native elements in Phrygian culture betoken a persistence of an ancient local civilization, not an Iron Age intrusion from Thrace.

Of an invasion from Europe we have indeed unambiguous evidence from Troy itself. But it is to be ascribed to the Treres; a band of barbarians settled on the ruins of Homeric Troy and introduced to the Troad a new mode of decorating the local pottery —by the application of big horn-like knobs—a style that was evolved in the Late Bronze and Early Iron Ages in Silesia and the adjacent lands and is best represented in the Lausitz pottery. That culture spread far into Russia, and the new settlers of Troy may have become acquainted with it there. In that case their identification with the Treres would be quite simple.

It is satisfactory to have found the Treres whom we met in the last chapter, but for the evidence of the western intercourse to which our traditions point we must evidently probe deeper into the mound of Troy. Phrygians were in Anatolia by the time of the Trojan War, but our Treres' village is built upon the ruins of the

[1] Körte, *Gordion.*

Mycenaean Troy which the Achaeans had sacked. In the Mycenaean town itself we find, of course, rare Mycenaean and Minoan vases ranging in date from the XVIth to the XIVth centuries, but such imports need not denote any colonization from the West; the bulk of the pottery is most closely allied to the Minyan of Greece. This may be due to a movement parallel to that which brought the same fabric to Hellas, but most authorities hold that its history is to be found in the five older towns upon the accumulated debris of which Homeric Troy was reared. It is in these deeper strata that the particular links between Asia Minor and Europe that concern us here must primarily be sought. And abundant evidence for cultural contacts with the West is indeed forthcoming in the lower levels, especially in the important town known as Troy II.[1] At the same time the pottery from this city is identical with that found in a small mound, Boz Euyuk,[2] in Phrygia proper, where the prototypes of the Phrygian vases from Gordion are also to be met.

Nevertheless there is some ambiguity in the links which unite Europe and north-west Asia Minor at this epoch. Several ceramic types are common to Troy II, Macedonia and Thessaly in the period of our proto-Hellenic culture. Other groups of objects, stone phalli, clay stamps used for painting the person, and certain types of stone and horn axes, recur both at Troy and in the Copper Age stations of Thrace described above. On the other hand the civilization of Troy is composite and the Asiatic inspiration is the most powerful. The truly Thracian painted pottery never crossed the Straits, and the most peculiarly European objects at Hissarlik are stone battle-axes.

Hence it must be confessed that the evidence for an invasion of the Troad from Europe is incomplete. We can only say that a cultural community subsisted between both sides of the Dardanelles somewhere about 2000 B.C. At a later date we should seek in vain for the same degree of unity. The complete absence from both sides of the Straits of types corresponding to the Middle Bronze Age of Central Europe and the rarity of the correlative Early Iron Age forms are negative facts of capital importance; if the traditional connections between north-western Asia Minor and the Balkan lands are to be upheld by archaeological means we must rely on evidence from an earlier period. It is striking that the context

[1] For Troy II, see Childe, *Dawn*, chap. iv.
[2] *A.M.*, xxiv, pp. 6 ff.

in which such evidence is forthcoming is the same as that in which the roots of Hellenic culture were detected further west. The reader can now see that the suggestion that the Trojans were Greeks of a sort actually gives support to our theory of the origin of the Hellenes.

3. *Ligurians and Italici*

As in other Mediterranean lands, non-Indo-European peoples have left memorials of their presence in the form of place-names in the Apennine peninsula. To them may be attributed the cultures [1] both of South and North Italy created by men of Mediterranean race—Siculi in the South, Ligures in the North. Moreover, Aryan Illyrians were in historical times settled in South Italy. It is not impossible that some of the Copper or Bronze Age material of Apulia which exhibits a certain parallelism to the culture identified in the eastern Balkans about 2300 B.C. as proto-Hellenic may belong to Illyrians. That, however, is a very debatable question which cannot be discussed here. We are concerned with the ancestors of the Umbrians, Oscans and Romans who, thanks to the tenacious conservativism of the latter, may be identified with some degree of certainty.

The Italic dialects [2] of historic times fall into two groups distinguished by the treatment of the Indo-European labio-velar sound k^{u}. The Latins and Faliscans in Central Italy preserved this sound as *qu* and are therefore termed Q-Italici, the Oscans to the south and east and the Umbrians further north labialized k^{u} representing it by *p* and are known as P-Italici for this reason. Apart from this phonetic cleavage the Italic dialects are united into a single linguistic family by many deep-seated bonds of kinship although they share many phonetic and grammatical peculiarities with Celtic. Moreover, several social, political and religious institutions, common to Latins, Oscans and Umbrians, may well be relics of their coexistence as a single people in prehistoric times.

These tribes do not become truly historical before the fifth century B.C. But thanks to Roman conservativism their ancestors are traceable by archaeological evidence nearly a thousand years earlier. In the XVth century B.C. a new people made their appearance in the Po valley among the old Mediterraneans of Upper Italy. Unlike their Ligurian predecessors and neighbours, the

[1] For these see my *Dawn*, chaps. vi and xvii.
[2] Conway, *The Italic Dialects*.

intruders cremated [1] their dead, depositing the ashes in cinerary urns which were laid out, closely packed together, in two cemeteries near each village. The villages themselves were pile-structures on the dry land and are known to archaeologists as *terremare.* They were always laid out in accordance with a deliberate plan. The latter reproduces to the smallest detail the Roman camp of historical times: the settlement was surrounded with a moat (corresponding to the *fossa* in the Roman *castra*), and a rampart (the *vallum*) was traversed by two main roads intersecting at right angles (the *cardo* and *decumanus*), while in the south-east quarter a low mound (the *arx*) itself girt with a moat, was heaped up, within which a sacrificial trench and pits were dug. The exact correspondence in plan between these structures and the Roman *castra* has led most serious students of Italian prehistory to identify their builders with the Italici and the ancestors of the later Romans.[2] And like the Romans the *terramaricoli* (to use a convenient Italian name for the inhabitants of the *terremare*) are revealed as well organized, rigidly disciplined, pious and industrious husbandmen, pastoralists and metallurgists, and at the same time well equipped both for offence and defence with the javelin and the dirk and possessed of domestic horses.

The *terramaricoli* must have spread all over Italy during the XVth–XIVth centuries B.C., though the Mediterraneans were nowhere exterminated. A true *terramara* identical in plan with those of the Po valley was planted as far south as Taranto some time before the close of the Mycenaean period in Greece, and the pottery and bronzes exhumed from its ruins belong to North Italian types.[3] In the Late Bronze Age (roughly the XIIth century) fields of cinerary urns similar to those deposited in the northern cemeteries and containing bronze pins and other objects derived from *terremare* types were laid out at Timmari near Taranto and at Pianello in the Marche. The material from the latter site leads on to that revealed by Early Iron Age cemeteries on the Alban Hills, a region hallowed by Roman tradition.[4] And the Alban material

[1] Ridgeway's statement that the *terramaricoli* inhumed is in flat contradiction with a very large mass of evidence for early cremations. On this see Peet, *The Stone and Bronze Ages in Italy and Sicily.*

[2] So Helbig (*Die Italici in der Poebene*), Modestov (*Introduction à l'histoire romaine*), Pigorini and Colini (summarized by Peet), Peet himself, op. cit., and von Duhn, *Italische Gräberkunde.* Among the dissentients may be mentioned Brizio, de Michelis, Ridgeway (*Companion to Latin Studies*), and Randall MacIver.

[3] *Dawn*, fig. 49.

[4] *B.P.*, xxxv–xxxvi. Cf. Randall MacIver, *Villanovans and Early Etruscans*, pls. xvi–xix.

may find its continuation in the early graves from the Roman Forum itself.[1] This chain of cemeteries, taken in conjunction with the historical survivals alluded to above, makes the demonstration that the Romans were descended from the *terramaricoli* as perfect as any purely archaeological argument can well be.

But were the *terramaricoli* the ancestors of the Umbrians and Oscans, the P-Italici, too? Were they, that is, truly the proto-Italici or only the proto-Latini? That is more debatable. Helbig, Pigorini, Colini and Peet give an affirmative answer. In the Early Iron Age, Reggio Emilia and Tuscany were occupied by a people who cremated their dead and who are conveniently designated by the term Villanovans. It is practically certain that they were the Umbrians. Dr. Randall MacIver has recently shown that in Tuscany the cremation graves of the Villanovans were superseded after a time by inhumation interments, which he ascribes to the Etruscans. Now Pliny tells us that the Etruscans took three hundred cities from the Umbrians. The Villanovans whom the Etruscans displaced must then be the Umbrians. At the same time Peet and the Italian authorities just cited consider that the culture of the Villanovans, like that of the people buried in the Alban necropoles, was derived immediately from that of the *terramaricoli* and therefore that the Villanovans were the descendants of the Italici of the Po valley. As links they adduce two Late Bronze Age cremation necropoles at Bismantova and Fontanella respectively in North Italy.

Modestov, Randall MacIver and others contend on the contrary that the Villanovan civilization was due to a fresh wave of invaders coming from Central Europe. They have not indeed been able to put their fingers upon a prototype of the Villanova culture in Hungary or anywhere else. Yet I may inform them that a possible prototype for the characteristic Villanovan cinerary urn does exist in the Middle Bronze Age pottery of Hungary.[2] Mr. Harold Peake[3] has also drawn attention to the distribution of a certain type of leaf-shaped slashing sword which he holds was introduced into Italy by the P-Italici. Nevertheless I do not find the archaeological evidence decisive on this point; the issue between one or two invasions from Central Europe must depend

[1] But on the whole the Forum graves seem strictly contemporary with the Alban.
[2] *Dawn*, fig. 94.
[3] *Bronze Age*, p. 122.

partly upon our view of the time needed for the differentiation of proto-Italic into its Q and P branches.

Now, if the culture which Modestov regards as intrusive—that of the Villanovans—could be shown to belong to the P-Italici as a whole, his view would certainly be the most acceptable. But this is not the case. Peet[1] has admirably shown that the Early Iron Age civilizations of the Oscan regions—Picenum and Campania—are not derived from the Villanovan culture as such, not even from its earlier phases. Moreover, in these regions the funeral rite was not cremation but inhumation. If then the contemporary civilization of the Oscans was so different from that of their kindred the Umbrians, the difficulty of regarding the latter as fresh arrivals in the Early Iron Age becomes insurmountable.

However, the practice of inhumation among the inhabitants of Oscan territories raises fresh perplexities. To explain it von Duhn[2] has formulated the theory that both Oscans and Umbrians were a fresh wave of invaders who buried their dead and only reached Italy after the Villanovan culture had attained its apogee. Now, as we have seen, inhumation did in fact begin to take the place of cremation in part of the area occupied by the Villanovan culture proper. But we have agreed with Randall MacIver that these inhumations were due to the Etruscans. We cannot therefore accept von Duhn's hypothesis and must look elsewhere for the explanation of the Oscan burial rites.

I would suggest that the inhuming people of southern Italy were in the main descendants of the old Mediterranean stock who had inhumed their dead from neolithic times. For the conversion of the Picenes and Campanians into Oscans I would appeal to the Bronze Age invasion by Italici attested by Taranto, Timmari and Pianello which as we have just seen are the sites of settlements by the *terramaricoli*. At least in the sphere of metallurgy it is certain that the culture of the invaders from the north with their brooches and winged celts superseded the older culture which had previously been orientated to the south-east (the Early Iron Age cultures may be regarded as derived from this Middle and Late Bronze Age civilization). It may well have been the same with language. Yet the newcomers need not have constituted more than a conquering minority and may have become assimilated

[1] *B.S.R.*, iv.

[2] von Duhn, *Italische Gräberkunde*; he admits that the Romans and the Villanovans were descended from the *terramaricoli*. Both would be Q-Italici.

in racial type and burial rites to the subject population which was far less barbarous than the Ligures of Upper Italy.[1]

If this analysis be correct, if, that is, the civilizations of Umbria, Latium, Campania and the Picene coasts were all distinct by the beginning of the Iron Age, and can only be connected colaterally through a Bronze Age culture from which all were developed (on the hypothesis here adopted), it follows that the unity of the Italic language must be referred to the earlier date when a cultural unity also subsisted. Now that cultural unity was a reality in the Middle and Late Bronze Age when representatives of a single culture were scattered from one end of Italy to the other. But this common culture was that of the *terramaricoli*. Hence I would incline to see in the *terremare* of the Po valley the memorials of the undivided Italici, in the *terramara* of Taranto and the necropolis of Timmari some of the proto-Oscans, in the cemeteries of Pianello and the Alban Hills proto-Latins, in those of Fontanello and Bismantova proto-Umbrians. To the latter I would ascribe not only the Villanova culture of Etruria and Umbria but also the oldest graves at Este; the Illyrian Veneti would then be responsible only for the second phase of the Iron Age at the latter site.

Having then identified the proto-Italici in the Po valley, can we trace them further back into the past? The structure of the *terremare* suggests very forcibly that their builders were akin to the people who built pile-dwellings on the Alpine lakes in the late Stone Age. And there were lake-dwellings on the Italian lakes by the Copper Age and all through the Bronze Age. But the material from the *terremare* cannot be derived from this group nor yet from the Swiss. It points rather to Carniola, Croatia or Bosnia. In the latter region some lake-dwellings of the latest Bronze Age have yielded pottery almost exactly like that of the Italian *terremare*. On chronological grounds the Bosnian material cannot indeed be looked upon as the parent of the Italian but rather as a parallel development of one common stock. In some sense this common stock is in turn related to that Balkan culture which we were able to recognize as early as 2200 B.C., and very specially to the Early Iron Age civilization of Macedonia. On

[1] This view is quite in harmony with the results of W. R. Bryan's valuable study of the Early Iron Age in Latium, *Italic Hut Urns and Hut Urn Cemeteries* (*Papers of the American Academy at Rome*, vol. vi, 1925), esp. pp. 159–67.

the other hand there are threads which might serve to attach the *terremare* civilization more especially to Bavaria or again to Moravia and Galicia. From this side it would be vain to attempt to unravel the tangled skein. For that we must invoke the aid of another ally.

4. *The Peoples of the Sea and of the North in the Egyptian Records*

Is is impossible to leave the question of the Aryanization of the Mediterranean without making some reference to the foreign invaders coming from the North who are mentioned and depicted upon Egyptian monuments between the XVth and XIIth centuries.[1] The Pharaohs of the XIXth and XXth Dynasties had to repel from the shores and frontiers of their empire fierce invasions. The assailants betoken the intervention of a new racial element in the Mediterranean world. Their facial types are strange to the older monuments, and they brought with them a new armament. There is no doubt that the appearance of these invaders on the Egyptain coasts was due to disturbances on the northern shores of the Mediterranean; the later Pharaohs expressly state that Peoples of the Isles were restless. It is highly probable that this restlessness was the reflex of the intrusion of Indo-Europeans or fresh bands thereof from more continental regions. Yet the exact relation of these events to our problem is still obscure and the experts themselves are much divided over the interpretation of the Egyptian references.

The first of the "Northerners" to appear are the *Shardana* mentioned under the form *Shirdana*, in the Tell el-Amarna letters (about 1400 B.C.). In the same documents the name *Danuna* occurs, which recalls the Greek Danaeans but seems here to designate a tribe dwelling in Canaan, while *Sheklal* mentioned about the same time may be the same as the later *Shakalasha*. Early in the XIIIth century the Shardana again figure in the records, this time serving as mercenaries in the army of Rameses II. This contingent had been formed out of prisoners of war taken by Pharaoh on the western frontiers of the Delta in a previous campaign. They acted as Rameses' body-guard in the Syrian expedition of 1287 B.C., when the band of peoples from the Troad described on p. 64 were overthrown. Then in 1229 B.C.

[1] See Hall, *Oldest Civilization of Greece*, pp. 172 ff., and *C.A.H.*, ii, pp. 281–3. Moret, pp. 336–44.

PLATE IV

BRONZE STATUETTE FROM SARDINIA
(British Museum)

[*face p 72*

fresh bands of Shardana, now allied with *Shakalasha, Thuirsha, Akaiuasha, Lukki* and Libyans, were defeated by Merneptah on the western frontier of Egypt. Finally in 1192 B.C., Rameses III routed a coalition of invaders coming both by land and sea consisting of the *Pulesatha, Uashasha, Takrui*, and *Danauna*.

The exact identification of these peoples and the localization of their home-lands are much disputed. The last group are the

FIG. 9. An Attack by Peoples of the Sea repulsed by the Egyptians (Medinet Habu, 1192 B.C.).

least debatable. The Pulesatha are Cretans, as the Egyptian representations show clearly enough, Fig. 10. They ultimately settled as the Philistines in Palestine, whither they brought that metopic pottery already described. The Danauna again are here Danaeans, Greeks. Perhaps they represent scattered bands from Agamemnon's host returning from Troy, since the Odyssey speaks of piratical raids on Egypt as commonplace events of that period.

The Takrui and Uashasha are less certain. The former have been identified with Teucrians from the Troad. But Dr. Hall and others prefer to see in them another Cretan tribe perhaps the inhabitants of what is to-day Zakro. Their headdress is certainly the same as that of the Philistines, Plate VII. And the name Teukroi does not occur in Greek literature before Callinos, who further tells us that they came to the Troad from Crete. Finally the Uashasha have been regarded as the Oscans of Italy or as the Axians of Crete.

Our view of the latter will depend upon the origin assigned to the invaders repulsed by Merneptah. These are admitted to be the Achaeans, Tyrrhenians (Etruscans), Sardinians, Sicilians and Lycians. The point in dispute is whether the Etruscans,

FIG. 10. Head of a Philistine.

Sardinians and Sicilians reached Egypt from their seats in the west or were still on their way thither. On the one hand it is striking that they attacked Egypt from the west. It is, moreover, clear that the well-known bronze statuettes found in Sardinia, one of which is reproduced on Plate V, depict the same people as the antagonists and mercenaries of the Pharaohs. On the other hand, the best traditions connect the Etruscans with Anatolia. The Shardana might then be Sardians [1] from Lydia, the Shakalasha, men of Sagalassos in Pisidia. In that case we might imagine that these three tribes travelled by sea to attack Egypt, and that, thwarted in their designs by Merneptah, they then went on to occupy and give their names to Etruria, Sardinia and Sicily. That issue will be decided when the chronological context of the Sardinian bronzes is accurately determined. At the moment it is only

[1] But the native name of Sardis was Sfard.

PLATE V

1. SHAKALASHA

2. SHARDANA

[*face p. 74*

possible to say that these statuettes belong to a late phase of the local Bronze Age civilization of the island. But many of the roots of that civilization go back to the end of the IIIrd millennium, a time when Minoan inspiration certainly did reach Sardinia.[1] We may add that the Shardana's swords as represented on the Egyptian monuments and our bronze statuette and by actual specimens from Palestine (Fig. 11) were not designed for slashing like the weapons employed in Greece by the XIIIth century and in continental Europe even earlier. They might on the other hand be regarded as a development of the West European daggers, like Fig. 12, 3, common in the Sardinian Copper Age, under the influence of XVIth century Minoan rapiers.

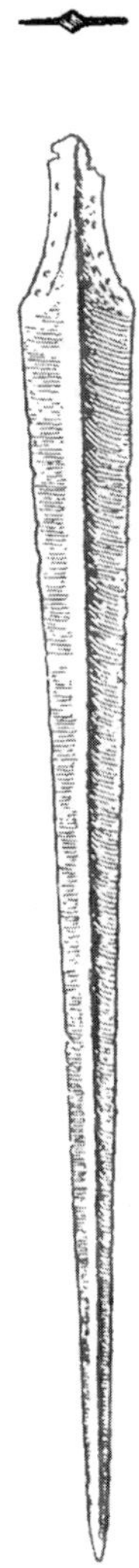

Fig. 11.
Sword of Shardana Type from Palestine (British Museum).

If the hypothesis of a western origin for the Shardana and their allies be adopted the exodus of the tribes from Italy and the adjacent isles might be ascribed to the pressure of Italic tribes marching southward; we have seen that the latter overran the whole peninsula in the XIVth century. Then the identification of the Uashasha with the Oscans might be accepted. We have already remarked that the ancestors of the Oscans had reached Taranto by about 1400 B.C. and some sort of intercourse with the eastern Mediterranean is suggested by the remains from their settlement.

In no case can it be said with certainty that any of the invaders (except perhaps the Achaeans [2]) enumerated above were Aryans. Indeed the leaders of Merneptah's

[1] *Dawn*, pp. 107 f. Cf. also R. Forrer in *Bull. de la Soc. Préhist. Française*, 1924.
[2] Sir Flinders Petrie does not agree that the *Akaiuasha* are Achaeans, but sees in them a tribe of the Syrtes region whose name may be recognized in Agbia, inland from Carthage, *History of Egypt*, iii, p. 112.

opponents had good Libyan names [1] and may be related to the modern Berbers. At the same time the type of the western assailants is depicted as blonde on the Egyptian monuments. The possibility that restless Aryans had mingled with the North African tribes is certainly one to be reckoned with.[2] Just the same remark applies to the Philistines. Though their Cretan origin is generally admitted and their faces are Minoan, the oldest examples of their feathered headdress (apart from those on the Phaestos disc of uncertain provenance) come from Mycenae. Professor Ridgeway long ago pointed out how the story of the "giant" Goliath might arise from the impression produced on the Hebrews by a tall European. If we are right in our dating of the Achaean invasion and in our views of the associations of the metope-style pottery a Hellenic infusion in these invaders of Palestine is not unlikely. And if the Hivites be in truth Achaeans the presence of some Aryans among the colonizers would be established.

The phenomena which we have just passed in review once more point to Central and Northern Europe. But it would be futile to plunge forthwith into the jungle of prehistoric cultures there distinguished by the archaeologists unless our path be lighted by the results of a kindred discipline.

APPENDIX TO CHAPTER III

The Achaean Period

Only the most summary indications can be given here of the evidence on which our view of the "Achaean period" is based. Archaeologically the epoch may be said to begin with the interruption of the regular commercial intercourse marked by the importation of Mycenaean vases into Egypt and the substitution of more warlike relations, such as those described in the Homeric Poems, indicated by stray slashing swords like Fig. 25, 3–4 (*B.S.A.*, xviii, pp. 282 ff.).

[1] On these see Bates, *The Eastern Libyans.*

[2] Dr. Christian holds that the blonde Libyans were Nordics who had come by way of Spain and introduced the "dolmens" into North Africa. He supposes that the same people proceeded eastward into Syria–Palestine, where they would emerge as Amorites and dolmen-builders. With the same racial drift he would connect the "predynastic" culture of Egypt which, contrary to the unanimous opinion of Egyptologists, he proposes to place *after* Dynasty VI (*M.A.G.W.*, lv, pp. 221 f.). The last proposition seems preposterous, and despite the undeniable similarity of the blondes depicted on XIIIth Dynasty monuments and much earlier representations, the first coloured figures belong to this late epoch. Cf. p. 102 below.

The turning point is dated by Mr. Forsdyke (*B.M. Cat. Vases*, I, i, p. xli) about 1250 B.C. Thereafter the pottery and the fibulae reveal a continuous evolution down to the full geometric period; with Mr. Forsdyke we may distinguish the following phases, which, of course, overlap :—

A. Late Mycenaean B pottery; fibulae as in Fig. 8, 1–4; bronze swords as Fig. 25, 3–4; iron rare; burials in chamber tombs, possible cremation at Muliana in Crete (*B.S.A.*, xiii); Mycenae and other citadels still occupied. Fibulae of types 1–2 are found at Mycenae, of type 3 at Mycenae, Kephallenia (Kavvadias, *Προϊστορικὴ Ἀρχαιολογία*, p. 367), Delphi (Homolle, *Fouilles de Delphe*, i, p. 7), Thebes (*Ἀρχ. Δελτ.*, 1917, pp. 151 ff.), and Vardino in Macedonia (*L.A.A.A.*, xii, p. 29), and of type 4 at Mouliana in Crete.

B. Sub-Mycenaean pottery (still including false-necked jars and stemmed goblets); fibulae like Fig. 8, 3–5; bronze and iron weapons; burials and cremations in chamber tombs—Vrokastro in Crete (*U. of Penns. Anthrop. Pubs.*, iii, 3), Salamis (*A.M.*, 1910, pp. 17 ff.), Assarlik in Caria (*J.H.S.*, viii, pp. 68 ff.), etc. The Tiryns hoard (p. 52) was probably put together during this phase. Pottery of this style, Wace's granary class, was in use during the last days of the citadel of Mycenae (*B.S.A.*, xxv, p. 40).

C. Proto-Geometric pottery without distinctively Mycenaean types. In North Greece partly contemporary with B in cist graves and chamber tombs containing also fibulae of types 4–5, iron ornaments and unburnt corpses (Wace and Thompson, *Prehistoric Thessaly*, pp. 209–15—Theotoku, Marmariani and Skyros); definitely later in the "pyres" of Halos in Phthia (Thessaly) and in the "bone-enclosure" at Vrokastro in Crete, containing fibulae such as Fig. 8, 6, iron swords and cremated bones.

We regard phase A as distinctively Achaean, and its pottery as the ultimate result of modifying native Mycenaean ware to suit the taste of the new ruling class whose accession to power, e.g., at Orchomenos, might have been anterior to the final establishment of the new style. The sub-Mycenaean pottery is essentially a continuation of the foregoing, and may be the product of the same society exhausted and disrupted by the Trojan War. Philistine pottery seems a parallel product (cf. Fig. 6). In southern Greece the proto-Geometric style is so closely allied to the above that an ethnic break cannot be asserted, though the local geometric styles that arose therefrom belong to the period of stabilization after the Dorian migration. In North Greece, as in Macedonia, proto-Geometric wares are older local fabrics modified by the influence of Mycenaean technique. Only in Crete, i.e., in the bone-enclosures of Vrokastro, does the appearance of a mature variety of this pottery mark such a clear break that we may connect it with the advent of the Dorians.

CHAPTER IV

PRIMITIVE ARYAN CULTURE RECONSTRUCTED BY LINGUISTIC PALAEONTOLOGY

In the last two chapters we beheld Aryan peoples emerging from the darkness of prehistory. In Hither Asia we believed that we could catch the first faint echoes of Indo-European speech on the tablelands of Iran by the begining of the second millennium B.C. By 1500 B.C. it was clear that the division into *satem* and *centum* languages was already established, and that an Indo-Iranian dialect not very far removed from Vedic Sanskrit was already being spoken. In Greece we thought that we could provisionally detect the Hellenes before the end of the IIIrd millennium, and in western Asia Minor we found it difficult to place the intrusion of the Phrygians very much later. Finally we recognized the Italici as a well-defined stock in Upper Italy by 1500 B.C. We must then conclude that the dispersion of the Aryans had begun by 2500 B.C.

But the Aryans we have identified appear as it were on the margin of history widely separated in space, their antecedents still shrouded in obscurity. In Hither Asia we have not succeeded in isolating any archaeological evidence beyond the introduction of the horse that went indubitably with Indo-European speech. In Europe, though specific cultures have been ascribed to the Hellenes and Italici, the roots of those cultures are manifold. The attempt to trace the Aryans inductively from their earliest stations in the arena of history leads us to a veritable labyrinth of complicated and intermingled cultures each with a long and intricate history of its own behind it. There is no single thread to guide us certainly out of the maze, but rather a multitude of strands intertwined and entangled and leading along divergent paths. To unravel the tangle we must have recourse to the deductive method, we must, that is, seek in the remoter and simpler phases of prehistoric civilization for a cultural group which may link up and gather together the loose ends of the skein.

In this quest the science of linguistic palaeontology offers to be our guide. This science claims to reconstruct the environment of the still undivided Aryan people and to conjure up the image of their spiritual and material culture. The words and names which recur in a plurality of the separate Indo-European languages, duly transformed in accordance with the phonetic laws as described on page 10, constitute in their totality the surviving vocabulary of the original Aryans. The objects and concepts denoted by those words are therefore the objects and concepts familiar to the ancestors of the Indo-European peoples. The sum of such corresponding terms would then depict the culture of the primitive people.

Certain reservations are, of course, necessary. In the first place the sum of surviving equations can only give a fragmentary picture, a sort of limiting outline, of the complete life lived by the Aryans. Through migrations, intermingling with other races, commercial relations with alien civilizations and the autonomous local growth and specialization of arts and cults, many words have been lost and replaced by others. Allowance must also be made for changes in the meaning of the word itself. Finally even strict compliance with the appropriate phonetic laws is not an infallible test of descent from the parent speech. The possibility is always present that the word in question came into the several languages by borrowing after the separation of their speakers, but at a date so early that the sound-shifts had not yet become operative. That might happen with especial ease in the case of the languages of the European Aryans who seem to have occupied from remote times closely contiguous territories, and may in some cases have been the heirs of a common pre-Aryan culture. For this reason many philologists since Fick have only accepted as belonging to the parent speech words found in Indo-Iranian on the one hand and in an European language on the other ; Schrader is content with words occurring in both *satem* and *centum* tongues. And, of course, the possibility of such borrowing infects especially the very cultural terms with which we are chiefly concerned.

The philological picture of Aryan civilization is then at best a minimum one. Attempts have been made to supplement and fill in its outlines by ethnographical methods. By comparing the customs, beliefs, institutions and industries of the several Indo-European peoples, it is hoped to isolate in the residuum common to all traits inherited from the period of coexistence. While useful in controlling linguistic data, I hold this attempt mistaken. The only Indo-European peoples of whom we have really early information, Indo-Iranians, Greeks and Romans, were as we know, intruders into an area of older civilizations of which they were in a measure the heirs and by which they were profoundly affected. Celts, Teutons, Slavs and Lithuanians only appear to us after they had been exposed for centuries to commercial penetration from the Mediterranean and inextricably mixed with one another, and, at least in some cases, with pre-Aryan populations who formed a common substratum in several areas. In each case it is extremely hazardous to say what in their civilization is due to pre-existing peoples, what to cultural borrowing, what to inheritance.[1] As a matter of fact the common culture deduced by this comparative method cannot claim to be specifically Aryan in the sense that the common traits observed among all the Bantu-speaking tribes are distinctively Bantu. It is so attenuated in character that it might belong to almost any primitive tribe in aboriginal Siberia or pre-Columbian America.

To illustrate the last point we may begin our study of Indo-European culture with religion. In this domain a comparison of customs and beliefs will only lead us to a nebulous complex which cannot serve to characterize a distinct society. The various schools of sociology will find in the results of such comparison reasons for applying to the Aryans their own pet explicative hypothesis. For instance, that mystic magical might, the Melanesian *mana*, may be discerned specialized and vaguely personified in the departmental gods (Sondergötter) of the Romans and Lithuanians or appropriated by a deity in Varuna's *māyā* or Odin's spells. The animist again will find material for a background of ancestral spirits in the cult of the dead so elaborately traced by Schrader.[2] That magic powers and ghosts played their part in the conceptual world of the Aryans, as among other peoples, must be at once admitted. But all that does not reveal anything distinctive.

But philology discloses, besides the background of magic and animism thus guessed at, more imposing and distinctive figures—real gods. The Aryans must indeed have worshipped more or less personified and individualized celestial beings whom they designated by a native word, *daevos*, the Bright Ones (1).[3] Among these there

[1] That is well illustrated by Frazer's argument for Aryan "matriarchy" from the usages of Mediterranean Aryans or Hirt's inference to a patriarchal monogamy which might at once be paralleled among the Babylonians or Assyrians.

[2] Article "Aryan Religion" in *Hastings' Enc. of Rel. and Ethics.* The cult was not conducted at the tombs, and that may imply that the spirit was not conceived as attached to the grave and the moral remains there enshrined as in Mediterranean religion.

[3] Numbers refer to the list of equations at the end of the chapter.

was at least one who stood out in specifically human form, *Dyeus pater*, the Sky-Father (2). He must have been to some extent a tribal or even a national god. He retains this rôle among the Greeks and Romans ; in the East his place has been usurped by Varuna even in pre-Vedic times just as in the Veda itself Varuna is losing ground to the warrior Indra and in the post-Vedic age Indra in turn is effaced by Vishnu and Rudra. Traces of other deities are less certain. Varuna may be related to *Οὐρανός*, and the inversion of the rôles of this deity and Dyeus as between India and Greece may indicate an ancient rival of the Sky-Father. Warlike Storm Gods (*Marutaḥ* and *Mavors*) and a fair Lady of the Dawn (*Uṣas 'Hώς*) are at best attenuated and hypothetical figures. Heavenly Twins (*Aśvinau, Dioscuri*), connected at once with horsemanship and navigation, have such extraordinarily similar traits in Vedic psalms, in the Homeric Hymns, and in archaic Lithuanian verses that their cult in the primeval period seems to the writer a likely conjecture. On the other hand no Earth Goddess, spouse and counterpart of the Sky Father, is traceable in language.

The deities certainly disclosed are important enough. If there be any truth in the speculations of Durkheim, Frazer, Perry, and other sociologists, the personified Sky Father is the mark of a relatively advanced stage of intellectual development. Whether he was evolved out of a departmental spirit or the ghost of a deified ancestor or the worship of a culture hero or borrowed along with other elements of civilization from Mesopotamia, he remains an imposing and distinctive figure. And the results of sociology suggest that his sovereignty reflects some sort of political unity among the undivided Aryans who worshipped him.

As a matter of fact, the social structure of the Aryans likewise seems to betoken a certain cultural evolution. A very large number of sociologists contend that the system of reckoning descent through the female has everywhere and always preceded the more familiar patrilinear system. Of such uterine kinship the Indo-European languages reveal no trace ; the Aryan names for kindred (3) are exceptionally widely diffused and preserve a remarkable uniformity of meaning in all the linguistic groups. They all without exception refer to agnatic relationships. We are then warranted in inferring that the Aryan family was patrilinear and patriarchal. Probably in the light of the expression for "brothers' wives" it was a large unit, an aggregate of several generations living together under the rule of the eldest male ancestor as "house-father" (4) either under one roof, as in the communal household (*zadruga*) of the Balkan Slavs,[1] or as a movable group such as the Biblical patriarchs ruled.

For more comprehensive groupings we lack precise evidence. A set of words from the root **uik* "to enter" (5), varying in meaning from "clan" to "village" or "district", may indicate some sort of local organization, perhaps territorial clans grown out of the great family; while incompatible with pure nomadism, it is not obvious whether the tie attaching such a group to a district was agricultural settlement or just the possession of common pastures. Exactly what lies behind words like **sebh-* and **genos*—perhaps "sib" and "tribe"—is debatable. Nor is it certain that, above the patriarch of the agnatic family or clan, any tribal or national chief was recognized as the earthly counterpart of the tribal or national god. There was, however, a root **reǵ*, different derivatives of which came to denote "king" in Indian, Italic and Celtic (7). Thus Aryan religion and society correspond to a phase of development which may indeed be lowly in the scale but is not strictly primitive, as it presupposes a certain history behind it.

When we turn to material culture, these inferences receive confirmation and the field of our quest becomes further limited in time. It is at once clear that the Aryans had passed beyond the Old Stone Age or palaeolithic phase of culture. So it is unnecessary to pursue our inquiry into those remote ages which preceded the geological present. In the Old Stone Age, which covers the quaternary epoch and in Europe closed about the time when the glaciers had finally retreated from France and North Germany, men were just food-gatherers. Domesticated animals and agriculture appear first in the New Stone Age or neolithic phase.

Now it is certain that the Aryans possessed domestic animals. Not only are the names for dogs, cattle, sheep and horses and perhaps also goats, swine, ducks and geese common to very many Indo-European languages, but words for "gelding" and distinct terms for males and females indicate an acquaintance with the operations of cattle-breeding (8). There are also words for butter and perhaps for milking, but not, curiously enough, for milk (8b). It is moreover clear from comparative ethnography that cattle played a prominent part in Aryan economy. Among the Vedic Indians, the Iranians of the Avesta, the Homeric Greeks, the Romans, Celts, Teutons and Slavs cattle were the principal source of wealth.

[1] So Hirt, p. 705.

Philology itself reveals the use of cattle as a standard of value among the Romans and Anglo-Saxons (*pecunia, feoh*) and among the early Hindus the word for " battle ", *gaviṣṭi*, means literally " struggle for kine ". And in Zoroaster's hymns the Spirit of the Kine personifies Aryan righteousness in the dialogue. The same sort of argument induces the belief that the Aryans had domesticated the horse, which they named " the swift one ". The horse is indeed the Aryan animal par excellence in the early history of Mesopotamia, in the Veda and in Homer ; in Iran Darius boasts of having made his land " rich in horses " (*'uvāspa*) even before he mentions " rich in men " (*'uvamartiya*). The words seem also to have had an Indo-European feminine (*aśvā, equa, aszwà*), and Feist notes how often Aryan personal names in India, Iran, Greece and Gaul contain " horse " as an element. On the other hand the domestication of the pig is denied by Schrader ; the Indian and Iranian words in the older sources denote only the wild boar.

That brings us to the question of agriculture. In contrast to the developed pastoral terminology of all Indo-European languages agricultural equations common to the Asiatic and European branches are rare. There is a word for some sort of grain and perhaps for " plough " and " furrow ", while a common root came in both areas to be specialized to denote the grinding or milling of grains (9). At the same time according to Schrader the Aryans only recognized three seasons—a cold period, winter, a spring and a hot summer—but had no name for the harvest time, autumn. As against the paucity of European-Asiatic equations, however, there is a rich terminology both for the operations of tillage and for varieties of cultivated plants shared by the European languages of both the *centum* and *satem* branches.

The interpretation of these phenomena is disputed. Partisans of an Asiatic or a South Russian cradle for the race consider that the undivided Aryans were semi-nomadic pastoralists who only occasionally stooped to cultivate the soil by rude and primitive methods (garden culture) ; the advance to regular agriculture would, they suppose, have been first made in the Ukraine or Central Europe after the Indo-Iranians had separated from the parent stem. It would also be possible, and, I think, better, to argue that in this case many of the agricultural terms were taken over by the first Aryan intruders from a race of peasants whom we shall find occupying the Balkans and all Central Europe as far north as Magdeburg in Saxony in the New Stone Age.

On the other hand it is now possible for the advocates of an European cradle to contend that among the primitive Aryans agriculture was as important as pastoralism. The old conception that a phase of nomad pastoralism intervened between the pure food-gathering stage of hunting and fishing and that of settled agricultural life is no longer tenable.[1] In some cases at least, judging from the results of the excavations at Anau in Turkestan, tillage preceded stock-raising. Some authorities, such as W. J. Perry, go so far as to say that pastoral nomadism is everywhere posterior to sedentary agriculture and was only adopted by cultivators under the pressure of adverse climatic conditions or political convulsions. It may then be argued that the Indo-Iranians, ejected from the agricultural regions of Europe and thrust on to the Eurasiatic steppe, had just lost the primitive Aryan agricultural terminology in a period of enforced nomadism. I do not personally believe that any one-sided priority of one regime over the other can be regarded as a historical fact nor that it is possible to deduce *a priori* whether the Aryans were primarily pastoralists or peasants. Some sort of cultivation of the soil must in any case be admitted; at the same time the wide distribution of Indo-European speech as well as the habits of some of its users implies at least a phase of nomadism, but not of the extreme type observed among the Mongols of Upper Asia. In my opinion the state of things observed among many of the cow-keeping tribes of the Sudan and other parts of Africa approximates most closely to the primitive Aryan economy.

Besides these sources of food supply which they themselves controlled and which mark them as already "neolithic", it may be assumed that the undivided Aryans still resorted to more primitive pursuits such as hunting. Yet there is no Indo-European terminology for the chase. And the absence of words for fish may well indicate that the Aryans did not supplement their diet from the denizens of seas or rivers; for fishing is never mentioned either in the Veda or the Avesta, and the repugnance felt by the Homeric Greeks for a fish diet is notorious. Nevertheless, one fish-name is common to Europe and Asia, for the Tocharian word for "fish", *lakṣ*, is the same as the Old High German *laks*—Lith. *laszisza*, "salmon." It is also curious that no word for salt is common

[1] Good accounts of agricultural and nomadic life will be found in J. L. Myres, *The Dawn of History*, and Dudley Buxton, *Primitive Labour*. Cf. also W. J. Perry, *The Growth of Civilization*.

to Indo-Iranian and the European languages. Yet the latter and Tocharian know a term for that substance, **sel*. Finally the Aryans enjoyed a drink, **medhu* (10), made from honey, though no word for "bee" has survived.

Not only does a regular food-producing economy stamp the Aryans as neolithic, but they had gone further and were acquainted with at least one metal. Copper is represented by two terms, **ayos* and **roudhos* (11), though, as both words are probably borrowings, Kossinna holds that their use does not go back to the period of co-existence. Feist believes that gold and silver were also known in the primeval period; in any case words derived from the same roots, **gher* or **ghel* "yellow" and **reǵ* "shining", were at a very early date used to denote the precious metals. But though the Aryans knew metal and no doubt metal implements, it was probably rare and not worked locally, but imported. On the one hand, there is no Indo-European terminology for metallurgy; on the other, the names of certain artifacts are proper to a period when stone was still used for tools and weapons. For instance, the Teutonic **sahsaz* "a cutting weapon" (preserved in O.H.G. *mezzirahs* "blade"), comes from the same root as the Latin *saxum* "stone". Again the meaning of **akmon* fluctuates between a metal and a stone weapon (Lith. *asmũo* "blade", Sans. *áśman* "stone", "bolt", Greek ἄκμων "anvil"). Thus the Aryans were still in a stage of transition from the use of stone to that of metal, what archaeologists call the chalcolithic phase, at the time of their separation. This is a most important point for the prehistorian even though the succession of Neolithic, Chalcolithic, Bronze and Iron Ages, cannot be regarded as an universally valid chronological sequence.

The Aryan names for tools and weapons (14), objects with which the prehistoric archaeologist is particularly concerned, confirm the foregoing conclusion. The best attested implements are the awl and the razor. The Aryan weapons were the club or mace, the sling, the bow, the spear or pike, the knife-dagger, and the axe. Two points only need special notice: the large number of equations peculiar to Greek and Indo-Iranian, and, secondly, the interchange of meaning between spear and sword in the case of one word **karu*; the latter means that the Aryan sword was not originally a slashing but a thrusting weapon, very likely that particular type of pointed blade of stone or copper so common in the chalcolithic period which would serve equally well as a dagger or a spear-head according to the length of the handle to which it was fastened.

That the Aryans made pottery vessels might be concluded from the above results even without the equations cited by Schrader,[1] but naturally no shapes can be inferred from the linguistic data. Wood, too, played a prominent rôle in their industry. Indeed, the only craft for which an Indo-European designation has survived is that of the carpenter (15). One of his products for which a detailed terminology is preserved was a wheeled vehicle (16). But, as far as philology is concerned, this may have been anything from the wagon-dwelling of the nomad to the horse-drawn war chariot associated with the earliest Aryans in Mesopotamia and so characteristic of the Vedic Indians and Homer's Achaeans. The carpenter must have been also called upon to build the boats named in a very large number of languages (17). But here again, though a word for oar survives, the meaning of **naus* may range from the dug-out canoe used on a river to regular sea-going craft.

Weaving may be denoted by a group of terms from the roots **vi*, **vebh*, since words for "wool", and also perhaps "spindle", are traceable (18).

The sort of house inhabited by the Aryans would be of great interest could it be reconstructed. Equations exist for door, door-frame or porch and pillar as well as the whole structure (19). They certainly suggest something more substantial than the nomads' tent—even perhaps a porched house like Plate VIII, 1, the prototype of the Achaean megaron—but nevertheless might be applicable to such an abode. From the series *śála*, *cella*, *höll*, Schrader infers some sort of pit-dwelling such as is common all over prehistoric Europe. Walls of wattle and daub seem to Feist to be indicated by a group of words derived from the root **digh* "to smear", but these may refer rather to defensive earthworks. Terms exist, if not for "village" or "city", at least for some sort of strong places or refuges defended by ramparts (20).

We have then to search for a people who were no longer just food-gatherers or even pure nomad herdsmen, but who had already made distinct progress in the arts as in political organization and religious belief. It would be a considerable help in our quest if it could be shown that their progress had been partly at least inspired by one of the great civilizations which arose in the Ancient

[1] *Reallexikon*, s.v., Gefass.

East at a very remote date, and from which, according to a widely held belief, all higher culture emanated. This seems, indeed, to be possible. The names for metal give us one key. The word **ayos* may be derived from *Alasya*, the ancient name for the copper land of Cyprus.[1] In that case it indicates that the influence of the great prehistoric civilizations of the Aegean which have left a deep mark on the culture of the whole of Europe had reached the undivided Aryans. But their indebtedness to the civilizations of Mesopotamia was much greater; not only is the other Indo-European word for copper **roudhos* derived from the Sumerian *urud*(*u*),[2] but the Indo-European words for "ox",[2] "steer",[3] "star",[2] and "axe",[2] seem all to be of Sumero-Akkadian origin. Of course, in the case of such loan words the possibility of borrowing after the separation of the peoples must be kept especially in view; in the light of the distinctively Mesopotamian culture recently disclosed in India it may be that Hindus and Hellenes each borrowed independently such a word as *pilakku* on reaching the Indus and the Aegean respectively. However, I feel that the concordances are too numerous and too deep-seated to be thus explained away. I believe that the Aryans received their initiation into stock-breeding and metallurgy and perhaps some elements of their celestial religion directly or indirectly from the bearers of that great civilization which was flourishing in Mesopotamia by the IVth millennium before our era. If true, this is a very important point both for the identification of Aryan remains and also for the delimitation of the Aryan cradleland; Mesopotamian influence, while dominant throughout Asia, cannot be distinctly traced in continental Europe further west than Russia. Conversely, it is crossed by Aegean influence only in the latter area and Anatolia.

Having drawn our all too vague picture of primitive Aryan culture, we shall now proceed to try to narrow down in space, as we

[1] Pokorny in *K.Z.*, xlix, p. 128. But it is now held that Alasya does not denote Cyprus but rather some part of the adjoining mainland.

[2] Ipsen, *Sumero-akkadische Lehnwörter in Indogermanischen* in *I.F.*, xli, p. 417. I–E. **roudhos* = Sum. *urud*; I–E. **guou* = Sum. *gu*(*d*); I–E. **ĕstēr* = S–A. *ishtar*; I–E. **peleku* = Ass. *pilakku*. This author tries to show that the modification in the pronunciation of Sumero-Akkadian *i*, presupposed in the above equations, was only realized under the First Dynasty of Babylon, and therefore that Aryan unity lasted till about 2000 B.C. However, the reading of early cuneiform does not seem sufficiently precise to warrant such a conclusion.

[3] Feist (p. 411) would connect the equation I–E. **stauros* = Ass. *šūru* with the ancient Mediterranean steer-cult, but it seems more natural to regard this as another instance of direct Mesopotamian influence. And, of course, the Mediterranean cult itself came from Mesopotamia.

have in time, our field of investigation by circumscribing the area where the material remains of the Aryans must be sought. The guides are the fauna, flora, climate and physiographical features of the cradle deducible from the extant Indo-European vocabulary.

The fauna of the cradle included, besides the domestic animals enumerated above, the wolf, the bear, the otter, the pole-cat, the mouse, the hare, the beaver, the quail, some bird of prey, and the snake, since the names of all these animals recur both in Indo-Iranian and European tongues. They do not tell us much. The horse indeed would seem to limit the possible regions to countries lying north of the Eurasiatic mountain axis; south of that the horse was a late comer, as we have seen, while the typical draft animal was the ass, for which there is no Indo-European name. Again, as the Aryan horse was "swift" (cf. *áśva*, *ἵππος* *equus*, and *āśu*, *ὠκυς*, *acer* "swift"), it seems more likely to have been either the steppe horse of Przybalski or the desert horse of Anau (*Equus caballus Pumpellyi*) than the stout German forest horse (*Equus cab. Nehringi* Duerst)[1] which would tend to move the cradle eastward. On the other hand, if the Aryans really came from Central Asia, they should have known the camel, since the American excavations in Turkestan disclosed remains of that animal in a quite early settlement. The absence of a word for lion (the name of that beast was borrowed by the Greeks from a Semitic people and transmitted by them to other European languages) is as unfavourable to Asia Minor as to Mesopotamia or Africa. Some animals require a wooded environment, others water, but the majority of the rest have such a wide range as to be useless for our purpose.

If the names for tortoise, salmon and eel, found only in European languages, be accepted as evidence of the Aryan fauna, important consequences would follow. Schrader used the tortoise (*χέλυς* = O.Sl. *zely*) to prove that the cradle must lie east of longitude 46°, but Professor Kossinna has pointed out that a tortoise shell has been found at Svaedborg, a very early prehistoric site in Denmark. On the other hand, the eel probably and the salmon certainly are not found in rivers flowing into the Black Sea.

As for the flora admired by the Aryans in their cradle, the data are still more exiguous. They had a name for tree, but, except perhaps for the pine, no special species of tree is designated by a

[1] See Duerst in Pumpelly, *Explorations in Turkestan* (Carnegie Publications, No. 73), ii, p. 431.

word common to the European and Asiatic groups. The former, however, agree in terms for beech, pine, sallow, alder, ash, hazel, elm and maple. Of these the beech has played a prominent rôle in the history of the quest of the Aryan cradle ; it does not grow to-day east of an imaginary line running from Konigsburg to the Crimea and extending thence to the Caucasus. Hence it has been inferred that the Indo-Europeans must have lived together west of that line. But apart from the general reservation made in respect of words only found in Europe, it remains a little uncertain when this frontier was established ; for the post-glacial forests of Europe seem to have advanced in several waves in a westerly direction. The same uncertainty attaches to the silver birch invoked by Professor Bender [1] to fix the cradle between the Vistula and the Niemen.

The climate of the cradle was severe, snow as well as rain being familiar phenomena, while the summer was hot. In a word the climate was continental. Such a climate reigns almost anywhere in the Eurasiatic continent north of the mountain axis and east of the Alps.

Finally the physiographical features of the cradle were not well marked. Rivers and streams were indeed common, as the fauna alone would tell. They seem in fact to have presented the chief obstacles to locomotion, for the variation in meaning of the word **pont-* from "path" to "ford" or "bridge" implies that the vital points on the routes frequented by the Aryans were river crossings. However, there is no certain word for sea common to Europe and Asia. Only in the former area is a term denoting sea or mere to be found in a plurality of languages. Nor, though the Aryans had a name for boat, are there general verbs for navigation. The root **per* "to cross" is frequently used in this connection, and Schrader adduces this circumstance to demonstrate that the Indo-European boat was only used for crossing streams. The same author denies that the Aryans had any mountains before their eyes and holds that *giri* (Sans. *giri*=O.Sl. *gore*) meant forest.

Finally we may note that early contact between the Finnic and Aryan peoples is an established fact. Some philologists, including Isaac Taylor and Kossinna, in fact believe that the Indo-European and Ugro-Finnish linguistic families are sprung from a common agglutinating stock. It is in any case certain that the Finno-Ugrians borrowed many words from Indo-European

[1] J. Bender, *The Home of the Indo-Europeans*, 1922, p. 33.

languages, beginning possibly (but not probably) with primitive Aryan and then assimilating Indo-Iranian, Slavonic and Teutonic vocables. Indo-European borrowings from Finnish, of which **medhu* has been cited as an example, are unproven. Since the Finnic cradle is even more hard to locate than the Aryan, the fact of early contact between the two peoples is of little practical use at the moment.

The scene of the undivided Aryans' life—a continental region traversed by rivers, sufficiently wooded to afford shelter to bears and beavers but open enough to nourish hares and swift horses and to permit of the unimpeded progress of vehicles—might be located almost anywhere in Eurasia save in the Mediterranean basin, the lowlands of Hither Asia or western Europe. None of the sites generally selected by philologists are excluded by our picture. These include Central Asia, Bactria, Armenia, Anatolia, South Russia, the Danube Valley, Lithuania, Germany, and Scandinavia. Yet all are open to certain more or less grave objections.

In Central Asia the camel was early known to man, while it is said that the honey bee is absent. If the Aryans had originated in Asia Minor we should expect to find some traces of them in cuneiform tablets of the IIIrd millennium, and they should have had a name for the lion. The pure pastoralism on which Schrader mainly bases his advocacy of the South Russian steppes appears exaggerated. At the same time he is probably mistaken in reading into prehistoric Russia the conditions of the present. In the IIIrd millennium the river valleys at least must have been quite sufficiently wooded to meet the requirements laid down for the cradle. But the absence of salmon is a real difficulty. The same objection applies to the Danube valley. Poland and Lithuania, in prehistoric times marshy or densely wooded, are much less attractive when viewed in the light of archaeological data than they appear from a study of modern geographical handbooks. Scandinavia on the other hand looks much less unlikely in the same light. Still the North European forest horse was slow and heavy, and life in those regions was very largely based on fishing and maritime enterprise. Above all, as de Morgan [1] points out, these lands are the sources of amber, while the Aryans had no name for that precious gum.

[1] *Prehistoric Man*, p. 272.

Objections may therefore be taken to all the proposed identifications. We will therefore proceed to survey each region in turn in the hopes of finding in one of them a culture resembling that sketched above, and a people whose diffusion to the appropriate regions of Europe and Asia can be traced by archaeological methods.

APPENDIX TO CHAPTER IV

The more important equations relied upon for establishing the primitive culture of the Aryans are summarized below, the numbers corresponding to the references in the text. As in the rest of this chapter I rely principally on O. Schrader, *Prehistoric Antiquities of the Aryan Peoples*, and *Reallexikon der indogermanischen Altertumskunde*, 1st ed., 1902, 2nd in progress.

	Sanskrit.	Greek.	Latin.	Celtic.	Teuton.	Lithuan.	Tochar.	Armen.
(1) god	*devá*	—	*deus*	*dia*	O.N. *tivar*	*dìẽwas*	—	—
(2)	*Dyaúṣ*	*Ζεύς*	*Jupiter*	—	*Tiu*	—	—	—
(3) father	*pitár*	*πατήρ*	*pater*	*athir*	*fadar*	—	*pātär*	*hair*
mother	*mātár*	*μήτηρ*	*mater*	*mathir*	O.H.G. *muotar*	O.Sl. *mati*	*māṭaṛ*	*mair*
son	*sūnú*	*υἱος*	—	—	*sunus*	*sunùs*	*soyä*	—
daughter	*duhitár*	*θυγάτηρ*	—	—	*dauhtar*	*duhtė́*	*tkacer* —	*dustr*
brother	*bhrā́tar*	—	*frater*	*brathir*	*broþar*	O.Sl. *bratrŭ*	*procer*	*bair*
sister	*svásar*	—	*soror*	*siur*	*svistar*	O.Sl. *sestra*	—	*kóir*
father's brother	*pitṛvya*	*πάτρως*	*patruus*	—	A.S. *faedera*	—	—	—
grandson or nephew *	*nápāt*	*νέποδες*	*nepot-*	*nia*	A.S. *nefa*	O.Sl. *netiji*	—	—
son-in-law	*jā́mātar*	—	—	—	—	*żéntas*	—	—
daughter-in-law	*snūṣā́*	*νυός*	*nurus*	—	O.H.G. *snura*	O.Sl. *snŭcha*	—	*nu*
father-in-law	*śváśura*	*ἑκυρός*	*socer*	Corn. *hveger*	*svaihra*	*szészuras*	—	*skesrair*
mother-in-law	*śvaśrū́*	*ἑκυρά*	*socrus*	Corn. *hvigeren*	*svaihro*	O.Sl. *svekry*	—	*skesur*
husband's brother	*devár*	*δαήρ*	*levir*	—	A.S. *tacor*	*dė̀weris* *jentere*	—	*taigr*
husband's brother's wives	*yātaras*	*εἰνάτερες*	*janitrices*	—	—	Lett.	—	—
husband	*páti*	*πόσις*	—	—	*-faþs*	*pats*	—	—
woman	*jnā́*	*γυνή*	—	*ben*	*qino*	O.Sl. *žena*	—	—
widow	*vidhávā*	—	*vidua*	*fedb*	**viduvo*	*vĭdova*	—	—

	Sanskrit.	Greek.	Latin.	Celtic.	Teuton.	Lithuan.	Tochar.	Armen.
(4) house-father	*dáṁpati*	? δεσπότης	—	—	—	—	—	—
(5) clan, village	*viś*	Ϝικ*	*vicus*	*fich*	*veihs*	O.Sl. *vĭsĕ*	—	—
headman	*viśpáti*	—	—	—	—	*wiẽszpats*		
(6) ? sib	*sabhā́*	—	—	—	*sibja*	—	—	—
? tribe, clan	*jánas*	γένος	*genus*	—	O.H.G. *chunni*	—	—	—
(7) king	*rā́jan*	—	*rex*	*rí*	—	—	—	—
(8) dog	*śvā́(n)*	κύων	*canis*	*cú*	*hunds*	*szũ*	—	*sun*
ox	*gó*	βοῦς	*bos*	*bó*	O.H.G. *chuo*	O.Sl. *govędo*	—	*kow*
sheep	*ávis*	ὄϊς	*ovis*	*ói*	O.H.G. *auwi*	*awìs*	—	—
goat	*ajá*	αἴξ	—	—	—	*ožỹs*	—	*aic*
horse	*áśva*	ἵππος	*equus*	*ech*	A.S. *ehu*	*aszwà*	*yakwe*	—
pig	*sūkará*	ὗς	*sus*	—	O.H.G. *su*	O.Sl. *svinija*	—	—
ox	*ukṣán*	—	—	Cym. *ych*	*auhsa*			
steer	Z. *staora*	ταῦρος	*taurus*	—	*stiur*	O.Sl. *turu*	—	—
cow	*vaśā́*	—	*vacca*	—	—	—	—	—
cow	*dhenu*	—	—	*dini*	—	—	—	—
gelding	*vádhri*	ἔθρις	—	—	—	—	—	—
cattle	*paśú*	—	*pecus*	—	*faihu*	—	—	—
(8b) cheese	Z. *tuirya*	τυρός	—	—	—	—	—	—
fat	*ajya*	—	*unguentum*	*imb*	O.H.G. *ancho*	O.Prus. *anctan*	(butter)	—
butter	*sarpíṣ*	ἔλπος	—	—	A.S. *sealf*	—	*ṣälypä*	—
(9) grain	*yáva*	ζεά	—	—	—	*jawaĩ*	—	—
grain, bread	*dhānā*	—	—	—	—	*dúona*	—	—
furrow	*karṣú*	τέλσον	—	—	—	—	—	—
plough	*vṛ́ka*	εὐλάκα	—	—	—	—	—	—
(10) mead	*mádhu*	μέθυ	—	*mid*	O.H.G. *metu*	O.Sl. *medŭ*	—	—
(11) copper {	*áyas*	—	*aes*	—	*aiz*	—	—	—
	lohá	—	*raudus*	—	O.N. *raudi*	O.Sl. *ruda*	—	*aroir*
(12) gold	*hiraṇyam*	—	*aurum*	*gull*	*gulþ*	O.Sl. *zlato*	*väs*	—
(13) silver	*rájatam*	ἄργυρος	*argentum*	*argat*	—	—	*ārkyant*	—
(14) razor	*kṣurám*	ξυρόν	—	—	—	—	—	—
awl	*ā́rā*	—	—	—	O.H.G. *ala*	*ýla*	—	—
sling-stone	*áśan*	ἄκων	—	—	—	—	—	—
bow-string	*jyā́*	βιός	—	—	—	—	—	—
arrow	*íṣu*	ἰός	—	—	—	—	—	—
javelin	*śastrám*	κέστρος	—	—	—	—	—	—
spear	*śáru*	—	—	—	*hairus* (sword)	—	—	—
sword	*así*	—	*ensis*	—	—	—	—	—
axe	*paraśú*	πέλεκυς	—	—	—	—	—	—
(15) carpenter	*tákṣan*	τέκτων	—	—	—	—	—	—
(16) chariot, wheel	*rátha*	—	*rota*	*roth*	O.H.G *rad*	*rãtas*	—	—

	Sanskrit.	Greek.	Latin.	Celtic.	Teuton.	Lithuan.	Tochar.	Armen.
wheel	*cakrám*	κύκλος	—	—	A.S. *hweol*	O.Sl. *kolo*	—	—
axle	*ákṣa*	ἄξων	*axis*	—	O.H.G. *ahsa*	*aszis*	—	—
nave	*nā́bhi*	—	—	—	A.S. *nafu*	O.Pruss. *nabis*	—	—
yoke	*yugám*	ζυγόν	*iugum*	C. *iou*	A.S. *yuk*	*jùngas*	—	—
(17) ship	*naús*	ναῦς	*navis*	*noi*	M.H.G. *naue*	—	—	*nav*
oar	*arítram*	ἐρετμός	—	—	O.H.G. *ruodar*	—	—	—
(19) house	*damá*	δόμος	*domus*	—	—	O.Sl. *domŭ*	—	—
	śā́lā	καλιά	*cella*	—	O.H.G. *höll*	—	—	—
door-frame	*ā́ta*	—	*antae*	—	O.Icel. *ǫnd* (porch)	—	—	—
door	*dvar*	—	*fores*	—	*daur*	*dùrys*	—	*durn*
pillar	*sthū́ṇa*	σταλλα	—	—	—	—	—	—
(20) earth walls	*dehí*	τεῖχος	Osc. *feihuss*	—	—	—	—	—

CHAPTER V

THE CASE FOR AN ASIATIC CRADLE OF THE ARYANS

The hypothesis of an Asiatic origin for the Aryan peoples is the most venerable but the least well documented. Indeed it belongs in part to that realm of anthropological mythology the roots of which go back to the Biblical story of the Tower of Babel. In that world of prescientific speculation all races were derived from Asia which was regarded as a vast reservoir of peoples, and it was assumed that all migrations followed the sun from East to West. To this extent the doctrine of an Asiatic cradle of the Aryans is only one of the unfounded generalizations which anthropology and archaeology have been combating for the last seventy years. We now know that the relations between Europe and Asia have not been so one-sided as our ancestors believed, and that culture and population flowed in both directions.

But the theory of an Asiatic cradle did not rest exclusively on prejudice. The supposed high antiquity of Sanskrit and its apparent linguistic purity were powerful arguments in the hands of the Orientalists and even led Schlegel (1808) to assert that the parent language itself originated in India and spread thence westward. A rather similar idea has cropped up in the writings of Sergi [1]; he supposes that the ancestors of the European Aryans were a brachycephalic stock originally inhabiting the region to the north of the Hindu Kush. There they would have learned the language of the Mediterranean Hindus and carried it with them into Europe. But modern philology can no longer regard Sanskrit as in all respects the purest representative of Indo-European speech. The fine state of preservation of the original inflection, due in part to the very early fixation of the language in a metrical literature, must indeed be admitted. But phonetically Sanskrit reflects the parent speech less faithfully than many European languages; for instance, "Aryan" must have distinguished between the vowel sounds ă, ĕ, and ŏ which in Sanskrit are all alike merged into ă. Again the

[1] *Gli Arî in Asia e in Europa.*

Indo-Iranian change of *k* to *s*, going back to the XVth century B.C., is an early example of phonetic decay indicating some physiological divergence from the parent stock in its users.

Even deeper was the impression produced upon the older philologists by the references to an *Airyanam vaējaṅh*, an Aryan homeland, in the Avesta of the Parsis. From the localization of this ill-defined centre of Iranian life in Bactria or Sogdiana it was an easy step to the identification of these districts with the cradle of the Indo-Europeans. Indeed, to Pott, Renan, Mommsen and Pictet the theory built on this foundation seemed an unimpeachable truth. That was, of course, partly the result of an illegitimate extension of the term Aryan to embrace all speakers of Indo-European tongues. But though we have for convenience retained the name in that sense in this book, we have stated at the outset that its use as a national appellation by the undivided people is unproven and indeed unlikely. As a racial designation it is peculiar to the Indo-Iranians. At the same time the most that the phrases in question, all in late sections of the Avesta, imply is a vague reminiscence of the migration of the tribe to which their authors belonged.

The case for a Central Asian cradle built up upon the Iranian documents is thus deprived of its basis. On the other hand, it was with justice remarked that the Aryan languages in Asia to-day are in a minority and stand out like isolated peaks in an ocean of Semitic, Asianic, Dravidian, Mongolian and Chinese tongues. And we have seen that the same relations held good in Hither Asia at the dawn of history. Even though at the beginning of our era Aryan languages were spoken over a vast tract extending from the Mediterranean to the frontiers of China, which has only been lost to them as a result of the Mohammedan and Turkish conquests, these languages were still almost exclusively merely dialects of Indo-Iranian as contrasted with the multiplicity of long-established Aryan tongues in Europe. Thus fifteen years ago the intrinsic probability that all the Indo-European languages were natives of Asia seemed but small.

The discovery of the *centum* Tocharian language in the Tarim basin has invalidated this sort of argumentation; it has recalled from the grave the old ghost of the Asiatic hypothesis and has endowed the Orientalists with renewed vigour. The simplest explanation of the presence of a *centum* language in Central Asia would be to regard it as a last survivor of an original Asiatic

Aryan stock. To identify a wandering of Aryans across Turkestan from Europe in a relatively late prehistoric period is frankly difficult. If we were right in regarding the Scyths as Mongols, it will follow that the tide of migration, which in historic times brought to Europe the Huns and the Turks, was flowing westward already in the VIIIth century B.C. It might have begun even earlier—do not many authors see "something Mongolian" in the Hittites? And then it would be easy to comprehend how that flood in its successive waves had wiped out the Aryans from Central Asia, swept them into Europe or hemmed them in to mountain valleys such as the Tarim basin. At the same time the revelation of the cyclic desiccation of Inner Asia has provided a motive for the great exodus of the nomads, perhaps for their very nomadism. Such desiccation might have begun the process of expulsion and isolation which the incursion of the Mongols completed. The world of Upper Asia is historically a blank till the last centuries before our era. We know not what languages it may have contained.

Finally the old catchword, *Ex oriente lux*, which has ever inspired the partisans of an Asiatic home of the Aryans, has at last begun to justify itself against the onslaughts of those who have made their watchword, *le mirage orientale, die Trugspiegelung der orientalischen Kultur*. But a reasoned and documented case using the latest discoveries for the illumination of our problem has not yet been put forward.

1. *The Alleged Brachycephalic Invasion*

The earlier investigators of the Aryan problem operated with the concept of race in the anthropological as opposed to the cultural sense. They relied upon physical characters for the identification of a human group which might have diffused Indo-European speech. Of course, the racial features which are most obvious to the layman, the colour of the eyes, the tint of the skin, the texture of the hair, are only very exceptionally available to guide us in the case of prehistoric men. For the racial classification [1] of our remote forerunners the ethnologist must perforce rely exclusively on the less perishable portion of the body—the skeleton, which under favourable conditions endures for thousands of years. From

[1] On this see Haddon, *The Races of Man*, 1924; Pittard, *The Races and History*, 1925.

the bones exhumed from prehistoric graves the stature and other attributes of ancient men can be reconstructed. But anthropologists lay most weight on the conformation of the skull which is held to preserve very persistent racial peculiarities. The most generally adopted criterion is the ratio of head-breadth to head-length which, when reduced to percentages, is called the *cephalic index*. Skulls in which the breadth is 80 per cent. or more of the length are termed *brachycephalic* or short-headed ; where the ratio is 75 per cent. or less the skull is classed as *dolichocephalic* or long-headed ; indices between 75 and 80 denote *mesaticephalic* skulls. It should be noted that anthropometrists are now feeling grave misgivings as to the value of the cephalic index alone as a test of race, and many, such as Sergi and Schliz, prefer to rely exclusively on the contour or other details of cranial conformation. In any case the length-breadth ratio by itself gives but a very rough classification. In the regions with which we are concerned the further division of the dolichocephals into Mediterraneans, generally short and dark, and Nordics, generally tall and fair, is also important.

We may pass over the early authors who imagined that Europe was entirely depopulated at the end of the Ice Age and that the neolithic civilization (page 82) was introduced into the void by a wholly new population come from Asia. It has long been established that remnants of the men of the Old Stone Age formed a considerable element in the post-glacial population of our continent. Whether any of these early races had come from Asia does not concern us, as the culture of the Aryans was not that of the palaeolithic phase but of the chalcolithic or neolithic.

Now it has been widely held that the New Stone Age in Europe was ushered in by the advent of a new anthropological type come from Asia. The intruders would be the " neolithic brachycephals ". In the neolithic period the short-heads do in fact appear rather like a wedge driven in between the short dolichocephals of the Mediterranean lands and the tall dolichocephals of the North. This apparently intrusive race has been claimed as Aryan, by Sergi and de Morgan [1] among others.

This simple identification is, however, no longer possible. The supposed intruders did not as a whole possess the civilization ascribed on philological grounds to the Aryans, but were still in the ruder stage of hunting and fishing without domestic animals. We are now acquainted with a considerable number of brachy-

[1] E.g. *Syria*, iv, pp. 28 f.

cephalic skulls from Spain, France, Belgium, Britain[1] and Germany which are pre-neolithic. That is to say, these brachycephals although post-glacial were still just food-gatherers and did not even polish stone or flint.[2] In the light of the results obtained in the last chapter they cannot have been Aryans. Nor is their Asiatic origin any longer undisputed. Bosch Gimpera thinks they may have come from North Africa across the Straits of Gibraltar with the palaeolithic people known as Capsians. The distribution of these early short-heads is in fact notably western. But what is more important, a brachycephalic skull has recently been found in a palaeolithic deposit at Solutré in France.[1] Thus it is no longer necessary to regard the neolithic brachycephals as intrusive nor to postulate an invasion to account for European short-headedness. The idea of using brachycephalism alone to establish a link between Europe and Asia is unworkable.

Equally unsound were the earlier attempts to supplement craniological by cultural data with the same end in view. In his classical work on the Formation of French Nation, de Mortillet admitted that the first short-heads in Europe were pre-Aryan, but assumed that a fresh immigration of the same Asiatic stock, bringing with them the art of metallurgy and the rite of cremation, introduced Indo-European speech to Europe. But here we can to-day see a triple fallacy. In the first place there is no coincidence between cremation and metallurgy. In Central Europe some cremation burials are still neolithic, while throughout the Early Bronze Age the prevailing rite from Britain to Crete was inhumation. Secondly it cannot be proved that the rite of cremation originated in Asia; even in India the oldest graves contain unburnt bodies. Still less can it be maintained that the European Bronze Age was a mere reflexion of the Asiatic. By 1700 B.C., when the use of bronze was regularly established in continental Europe, our ancestors had evolved a whole series of forms which have no parallels or prototypes in the East. In the Aegean, where the roots of the continental Bronze Age lie, the divergence of West from East can be traced back to the middle of the third millennium and became a superiority by the middle of the second.[3] The European Bronze Age is a native product, not brought ready made from Asia nor requiring an oriental invasion to explain it. Finally there is no

[1] Keith, *Antiquity of Man*, pp. 139 f. and pp. 91 ff. *L'Anthr.*, xxxv, p. 189.

[2] See my *Dawn*, pp. 1–12 and 318 with literature there cited.

[3] The use of bronze, an alloy of copper with 10 per cent. of tin, was not known in Asia before 1700 B.C., and seems to have been introduced from Europe!

noticeable increase in brachycephalic skulls in Bronze Age graves; long-heads are still predominant as in the neolithic period.

However, in the immediately preceding chalcolithic period of Central Europe a distinctly brachycephalic race had played an important part in preparing the foundations of the Bronze Age. This race, distinguished not only by craniological marks but also by a culture of its own, is known as the bell-beaker folk or the Prospectors [1]:

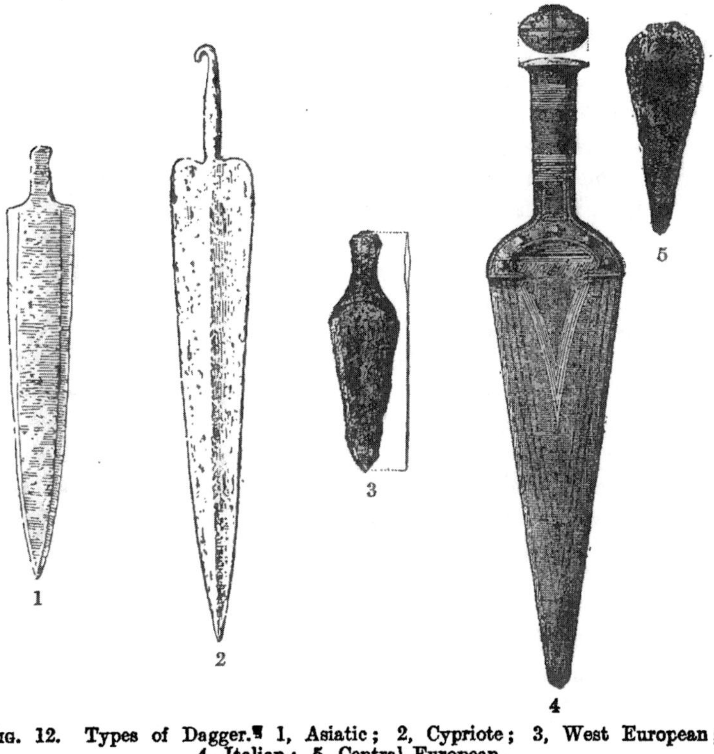

FIG. 12. Types of Dagger. 1, Asiatic; 2, Cypriote; 3, West European; 4, Italian; 5, Central European.

the former title is derived from the distinctive type of vase always buried in these graves, the latter from the fact that its bearers sought out ores and precious substances, while in Central Europe the first articles of value, gold and amber, are found in the same graves. But in continental Europe these intruders—they were not very numerous—did not come from the East but from the South-

[1] On the bell-beaker folk see my *Dawn*, pp. 121 f., 135, and 185 f.

west. The distribution of their graves—most numerous in South-west Germany, sporadic in Silesia and round Buda-Pest and non-existent further east—alone suffices to exclude the idea of an Asiatic immigration. But the grave-furniture is decisive. The most typical metal object is a very short, flat, triangular dagger with a broad tang widening to the blade without any distinct shoulder and probably inserted into a cleft wooden or bone hilt with a semicircular indent where it met the blade (Fig. 12, 3). This weapon is at once distinguishable from the Asiatic dagger with its pronounced shoulders and narrow tang, as illustrated by a growing series from Elam and Anau to Syria and Troy (Fig. 12, 1). On the other hand the Prospectors' dagger is very common in Western Europe and may ultimately have come thither from Egypt.

For there is in fact some evidence to indicate that these Prospectors did, in the last resort, come from the Eastern Mediterranean, though they did not reach Central Europe from that quarter. Both Peake and Giuffridi-Ruggieri hold that a type, which seems to correspond to our bell-beaker folk, originated in the Aegean, where a brachycephalic element is early found in the Cyclades and Crete. Starting thence, it is supposed that the Prospectors sailed westward through the Mediterranean and the Atlantic. It is certain that their physical and cultural types are found at an early date in Sicily, Sardinia and South France, but nowhere is the bell-beaker so richly or characteristically developed as in Central Spain, and it looks as if it was thence that the Prospectors diffused their vases and their daggers to Britanny, South France, North Italy and Central Europe. Be that as it may, it can hardly be contended that the brachycephalic Prospectors were the diffusers of the Aryan languages in Europe. In the first place they appear everywhere on the continent only in small numbers; they made no permanent settlements but, like the Arabs in Central Africa, were merely armed traders. They undoubtedly exercised a powerful civilizing influence, but are not likely to have changed the speech of the natives any more than the Arabs have imposed their language on all the negro tribes of Africa. And secondly the bell-beaker folk had their chief centres in just those parts of Europe where philologists and historians are disposed to recognize in historic times remains of pre-Indo-European languages, such as Iberian. In fact, even to-day in one centre of the bell-beaker culture, the Pyrenees, a non-Aryan language survives in Basque. The brachycephalism which still marks that people is already observable in the chalcolithic epoch, and may be due in part to our Prospectors. It may not be irrelevant to remark that the Basque word for copper is *urraida*, which may be connected with the Sumerian *urudu*, since Peake [1] has suggested that his Prospectors were Sumerians.

We have now assigned to its proper place in the formation of European metallurgy the brachycephalic element and at the same time excluded it from the Aryan race. It may here be convenient to mention the megalithic culture [2] since Peake attributes its diffusion to his Prospectors. The monuments in question, the huge stone graves known as dolmens, passage-graves and covered galleries, and the associated monolithic pillars and stone circles, do indeed afford one of the most conspicuous links between Europe and Asia—especially India. In Europe megalithic tombs are scattered all along the coasts of the North Sea and the Atlantic and on the shores of the Western Mediterranean to reappear in Bulgaria and on the Black Sea, whence they extend across the Caucasus into North Persia, while another group emerges in North Africa, Syria and Palestine and again, most significant of all, in South India and Assam. Most archaeologists consider that the idea of constructing these unwieldy tombs was diffused by a maritime race who set out from the Eastern Mediterranean in the search for metals and precious substances; for there is a rough coincidence between the distribution of the monuments and the substances in question. It is supposed that these early voyagers established trading stations or even dynasties where they found the objects of their quest and initiated the natives into their cult of the dead and the architecture which it inspired. In some form this view seems to me to be the right one, but none of its advocates have identified their treasure-seekers with Aryans. Siret calls them Phoenicians, Peake names them Prospectors and connects them with the Sumerians, Elliot Smith derives them from the Ancient Egyptians, and Perry, elaborating his views, considers them scions of Pharaoh's house, "Children of the Sun." Clearly then if the dolmen idea be of oriental origin, the navigators who diffused it cannot be the bearers of Indo-European speech.

There is, however, a school which holds that megalithic architecture originated in the North or West of Europe and spread thence

[1] *Bronze Age*, pp. 58 f. The latest discussion of the ethnology of the Iberian Peninsula will be found in *M.A.G.W.*, lv, pp. 110.

[2] On the megalithic culture see Perry, *The Growth of Civilization*, 1924, chaps. iv and v, and, for Europe, my *Dawn*, pp. 109 ff., 140, and 280. Peake, loc. cit.

eastwards. The advocates of a North European cradle for the Aryans might seize on this idea as a support for their thesis and a brief digression may be permitted here to examine their contentions. It is pointed out that the megalithic tombs of Scandinavia and Britain cover a purely neolithic furniture; in Spain and the Caucasus the tombs contain copper objects, while in North Africa and India the grave goods include iron implements. Moreover, some consider that the Scandinavian tombs are typologically the most primitive. So it is proposed to reverse the usual account of their diffusion and locate the original focus of dolmens in Denmark. Thence, it is suggested, tall sea-rovers with golden locks, the forerunners of the Vikings, set out in glorified dug-outs for Barbary and India. Wilke[1] has sought to buttress such a thesis by adducing ceramic parallels, and Christian[2] seems inclined to connect the blondes of Libya, known to the Egyptians and Herodotus, with dolmen-builders come from the icy North. Unfortunately Wilke's parallels are chosen haphazard from a mass of material disparate in origin and date and so carry no conviction, while Haddon[3] has noted that no dolmens occur where the blonde Kabyles are purest and most abundant. In any case, there are conclusive reasons against connecting the dolmen-builders, come they from North or South, with the Aryans. The distribution of megalithic monuments in Europe itself lies principally in territories which on the consensus of opinion were Aryanized only late—France, Britain and Spain. In North-west Africa and Palestine we know no Aryans, and finally in India the dolmens are located in precisely those parts which were last conquered by the Aryans; in the north dolmens are absent.

It results from the foregoing analysis that Asia's claim to the parentage of the Aryans or of the neolithic civilization in Europe cannot be established by an inspection of skull forms; the brachycephals as such are neither specifically Asiatic nor invariably Aryan. At the same time we have become acquainted with two groups of people very possibly of oriental (not specifically Asiatic) affinities who played an important part in the civilizing of Europe —the bell-beaker folk, who opened up regular trade routes through the interior of our continent, and the dolmen-builders whose maritime enterprise may have introduced the natives of its coasts

[1] *Megalith-kultur.*
[2] *Anthropos,* 1921–2, p. 583. Cf. p. 76 above.
[3] Op. cit., p. 36.

not only to the cult of the dead but to some at least of the arts of civilization. Neither of these peoples were either Aryan or natives of Central Asia. But the cultural material from the latter region which is rapidly accumulating provides the basis for a more plausible case than can be built up from mere cranial measurements.

2. *The Vase Painters*

At the very dawn of the food-producing era the shadowy but stately outlines of a mysterious civilization, majestic in its range, transcendental in significance for human progress, are to-day beginning to emerge from the morning mists that cover the scene of history as the last glaciers retreat. It appears from the Yellow Sea to the Adriatic as the first manifestation of men who had made the great advance from a food-gathering to a food-producing economy. The distinctive trait which holds together the far-flung ramifications of this primordial civilization is the art of vase-painting.[1] Beyond that few, if any, significant points of community can be isolated. The vase-painters indeed everywhere tilled the soil, but it is not clear that they all possessed domestic animals. At Anau, in Turkestan, for instance, the lowest stratum disclosed remains of cultivated plants, but the bones of domestic animals only made their appearance at higher levels. The vase-painters again generally polished stone, and almost certainly were acquainted with copper,[2] but distinctive types common to distant areas are lacking. The reader may then think that vase-painting is but a slender thread upon which to hang far-reaching historical conclusions.

But consider a moment what that art implies. To paint your clay with a permanent indelible colour which will not be destroyed but fixed by firing, that was a technique the secrets of which are not likely to have been twice discovered. One flash of genius in the brain of a nameless inventor made possible that art of which Attic vases and Doulton china are but elaborations. Nevertheless, I must insist at the outset that the painted pottery which concerns us is very far from being the same everywhere. From place to place the forms, technique, designs, the very aim of the artist,

[1] On the general question consult especially T. J. Arne in *Palæolontologia Sinica*, Series D, I, 2 (Geological Survey of China, 1925), H. Frankfort, *Studies in Early Pottery of the Near East* (R. Anthrop. Instit., Occasional Papers, 6, 1924), E. Pottier, in *Mémoires de la Délégation en Perse*, xiii, and R. Pumpelly, *Explorations in Turkestan.*

[2] This is no criterion of absolute date.

differ profoundly. And at all sites the painted fabrics appear before us tantalizingly perfect; we can only follow what is, aesthetically, a retrogression. Nevertheless, the diffusion of this magnificent art must, I think, denote a migration of culture if not of peoples. The immense range of its distribution in space is enough to account for very wide divergencies. Along latitude 40° our material is dispersed from longitudes 15° to 120°! (The sites are marked x on our map.)

First we meet painted sherds in the prehistoric midden-heaps

Fig. 13. Painted Vases from the Province of Honan, China. (After Andersson.)

of Japan. Then in China we have the newly discovered Yang Shao culture of the provinces of Honan and Chih-li, and further west round the head-waters of the Hwang-ho in the frontier districts of Kan-su (Fig. 13). Thereafter we must cross the now desert uplands of Chinese Turkestan to pick up the thread again in Transcaspia at Anau near Merv, in Khorassan, and on the Helmund, in Seistan. And finally, after an almost unbroken gap, we have another series of sites in Europe beginning on the Dniepr near Kiev and extending into Transylvania, Bulgaria, Thessaly and South

Italy.[1] At the same time south of the mountain axis painted wares are known from the Punjab,[2] Baluchistan,[3] Elam, the valleys of the Tigris and Euphrates, Cappadocia, Syria, Palestine and the Nile Valley. Such a distribution and its discontinuity are unintelligible on the existing geography of Asia and Europe. But in the six or seven thousand years which have intervened since that diffusion began, deserts and seas have contracted and expanded, forests advanced and retreated ; sites like Anau, now in the desert, or Petreny on the Bessarabian steppe were fringed with woodland when man settled there and hunted the wild boar. In fact, the authors of our culture seem to have skirted the grasslands fringing

FIG. 14. Painted Vases from Susa I.

the forest and shunned the open steppe as much as the desert. And in climatic changes the motives as well as the direction of these wanderings may be understood. The cyclic desiccation of Asia studied by Elworth Huntingdon [4] was a factor which induced early man still on the borderline between food-gathering and food-producing to roam from one end to the other of an as yet uncrowded world.

[1] Crete is deliberately excluded, since the oldest pottery there is unpainted and the new technique was probably introduced from Egypt, where it was much older, or Thessaly where the painted ware may partly be contemporary with the incised fabrics of neolithic Knossos.

[2] *Illustrated London News*, September 20, 1924. Plate VI here.

[3] *Arch. Survey of India*, 1904–5, pp. 105 ff., pl. xxxiii.

[4] *The Pulse of Asia*, 1907 ; cf. Myres in *C.A.H.*, i, pp. 6–86.

Nor is the painted pottery, even the oldest, in each area more restricted in temporal than in spatial range. Not only have a series of successive cultures associated with painted fabrics come to light at a given site or group of sites—two settlements at Susa in Elam, for example, four at Anau in Turkestan, two phases at Cucuteni in Roumania and in South Russia as a whole, two periods again in Thessaly. But further, the beginnings of such art are themselves far from synchronous in its several provinces. In Elam and Egypt vases were being painted by 5000 B.C.; in Thessaly and the Ukraine it seems unnecessary to go back much beyond 3000; the oldest Chinese and Transcaspian material is still really undatable.[1] Such chronological disparities are a salutary warning against hastily attributing the technique to any single and unmixed ethnic stock. In two thousand years much crossing and hybridization may have taken place. The process of diffusion may partly have taken the form of intertribal borrowing. Yet in those remote ages the world's population was smaller than to-day and even further from constituting a continuum over which cultural eddies might be freely propagated. Actual popular movements of the nature outlined in the last paragraph seem a necessary postulate to account for the dispersion of our material. It is therefore not illegitimate to inquire what racial element or elements assisted in such diffusion.

An answer to that question should materially help in the solution of the problem of the original focus of the art. The sites we have enumerated cluster in a striking manner around the great east-to-west mountain spine which divides the Eurasian land mass in twain. Now anthropologists consider that the same barrier separated the regions where two great branches of the human race were characterized: south of the axis the brown Eurafrican dolichocephals, north of it the Eurasiatic brachycephals.[2] The question just raised would then resolve itself into this: Were the first vase-painters Eurafrican or Eurasiatic? The former view is sustained by Elliot Smith and Perry[3] among others, the latter by Christian[4] and to some extent by Peake. The skeletal remains as yet available for study are hardly decisive. The skulls from

[1] Much higher dates have been assigned by others both to pre-dynastic Egypt and Elam (so Myres, loc. cit., cf. Moret, pp. 120, 200 ff.; Pumpelly, on the basis of a questionable geological and climatological postulate, dated Anau I about 8000 B.C. Professor Hubert Schmidt from an archæological standpoint not much earlier than 3000! Cf. the several articles in Pumpelly's book.

[2] Haddon, *Races*, pp. 142 ff.

[3] E.g. in *The Growth of Civilization*, pp. 24 f.

[4] *M.A.G.W.*, liv.

the graves found in China, India, Elam and South Mesopotamia have not been published. No adult burials are yet known from Turkestan, South Russia or Thessaly at the period in question, and for these areas we have to rely on stray skulls or on the skeletons of infants buried under the houses. The measurements of the extant material gives the following results: The infants from the oldest settlement at Anau,[1] the predynastic Egyptians and the vase-painters from Molfetta [2] in South Italy were all dolichocephalic of the type classed by Sergi as Mediterranean. One out of four skulls exhumed in Bulgaria,[2] belonging perhaps to a late phase of our culture, and probably one individual from Cucuteni in Roumania [2] belonged to the same stock. On the other hand, two of the Bulgarian skulls, one from Cucuteni and one from Levkas [2] in West Greece, were markedly brachycephalic. Finally

FIG. 15. Painted Vase from Susa II.

the two first-named sites and an older village in Thessaly yielded mesatocephalic crania, that from Bulgaria belonging to a man whose tallness may denote an infusion of Nordic blood.

These scanty data suffice at least to show that, whatever migrations did diffuse our material, they did not take place in an ethnological vacuum. The admixture which we envisaged above as a possibility appears as an actuality in Roumania and Bulgaria. The one common element in the skeletal remains would favour the view that the centre from which the ceramic art radiated lay south of the mountain axis and that its bearers were a branch of the Eurafrican race. That stock is still represented in India and Persia

[1] Sergi in Pumpelly, op. cit.
[2] Childe, *Dawn*, pp. 71, 87, and 318.

as in the Mediterranean lands proper. At the same time it must be remembered that there was a semi-negroid or negrito stock in Elam at the dawn of its history, and Dr. Christian [1] would assign to this element the bow-shaped knife found at Anau and Yang-Shao and still surviving in Malaysia and the Sudan. Still it does not necessarily follow that the Nile was the mother of the new invention of vase-painting. Egypt, lying on the western edge of the province, is far from its centre. Nor does it provide that balance between food-production and mere hunting which had to be postulated to explain the diffusion of the art. And finally the great divergence from site to site of the ceramic forms and technique and of the artifacts—and notably the absence outside the Nile valley of the quite distinctive predynastic pot-forms, the Egyptians' flint technique or their peculiar disc-shaped mace-head—precludes the idea of the immediate descent of all groups from the Nilotic.

On the existing distribution we should rather seek the first focus in Asia. Susa seems nearer to it. There Professor Myres' brilliant analysis discloses in the first village a band of hunters just settling down to agricultural life.[2] Yet the Susian fabrics and shapes cannot in themselves be taken for the prototypes of all the rest. We must still seek elsewhere, and it is worthy of mention that Professor Obermaier [3] can trace in India a typological series leading from the rough "hand-axe" used by palaeolithic man to the polished "neolithic" celt.

We are now in a position to face the question which alone could justify the inclusion of this lengthy disquisition in the present work: Was the first diffusion of vase-painting wholly or partly the work of Aryans? Undoubtedly the civilization just described, vague and attenuated as it is, is one of the most notable links between the Aryan lands of Asia and Europe. Yet as a whole the vase-painters cannot have been Aryans. *Qua* Ancient Egyptians, for instance, they belonged to other linguistic stocks. That is not, however, a final answer to our question. Although not as a whole Indo-European, the vase-painters may by foreign admixture or local differentiation have become Aryans, say in Upper Asia, and have entered Europe as such.

[1] Loc. cit., p. 61. His suggestion that the "Mediterranean" skulls from Anau belonged to prisoners captured by brachycephalic villagers is rather far-fetched.

[2] Cf. Frankfort, pp. 30-34.

[3] *Der Mensch der Vorzeit*, p. 331.

The Asiatic sites where painted pottery has turned up do indeed coincide rather closely with the earliest centres where Aryans appear. An Iranian dialect, Sogdian, was spoken in Kansu in the third century A.D.[1] and the same Chinese district is not far removed from the domain of Tocharian (page 8 above). We have already noted the significance of the new discoveries from the Punjab, and now we can stress the unpublished finds from the Helmund (? Hara'uvatiš) in Seistan where the *Airyanam vaêjaṅh* might be located. In Persia and Transcaspia the wares in question fall within the range of the earliest Iranian culture. Finally South Russia is by many associated with the first centre of European Aryans and we were led to locate the ancestors of the Hellenes in Thessaly (pages 50 and 59 above).

Not only so ; the vase-painters were, like the Aryans, possessed of copper and, at Anau, the same people appear before us as the domesticators of the Asiatic *Urus*, the Asiatic *Ovis vignei,* which will become the turbary sheep of " neolithic " Europe, and, most powerful argument of all, of the desert horse *Equus caballus Pumpellyi,* according to Duerst the first swift horse to be tamed and the ancester of the Bronze Age horses of Europe and Hither Asia.[2] These animals were among those known to the primitive Aryans, and they were very likely introduced into south-eastern Europe together with the arts of metallurgy and vase-painting by migrants from Central Asia. Thus the painted pottery appears as a significant link between Europe and those areas of Asia once occupied by Indo-European speech and associated, at least in Transcaspia and South Russia, with animals classed as Aryan.

It cannot, however, be assumed forthwith that our quest is ended. Having eliminated from our survey the painted wares of Egypt and Canaan we are very little nearer an unitary culture. The same divergences that were noted in the case of the vase-painting culture taken as a whole infect the remaining groups in Asia and Europe save that certain domestic animals are common to Europe and parts of Asia. As soon as we desert the abstract unity obtained by isolating and emphasizing the one fact of ceramic decoration and envisage instead concrete regional groups of cultures, the incoherence of the whole structure becomes glaringly

[1] Feist, p. 425. The Kansu pottery is not dealt with by Arne, but a preliminary publication by Andersson is given in *Ymer*, 1924, pp. 24 ff.

[2] Duerst in Pumpelly, vol. ii. The dog was probably not known at Anau I ; on the other hand, this animal and the horse are the only species likely to have been domesticated at Susa I.

manifest. With their individualization the several constituent sections of the whole tend to fall apart. At the same time the influence of extraneous cultures, omitted from the first survey, those of the Minoans of Crete or the historical Sumerians for instance, obtrudes itself as a disturbing factor. Without going into the intricacies of ceramic technique, so admirably handled in Mr. Frankfort's monograph, let us note a few simple points.

To begin with Europe,[1] the regions from the Dniepr to the Alt including Bulgaria do form an unitary province to which South Italy may with some reservations be attached, but the oldest culture of Thessaly resolutely refuses to be amalgamated with the North Balkan group. Nothing could be more different in forms, technique and ornament than the first neolithic pottery of North Greece and that of Transylvania. You may see the contrast in Thessaly itself when the true North Balkan culture and pottery do intrude into the eastern corner of that district in the second neolithic period. In the north the ware is thick and true handles are unknown; in the south the vases are very fine and equipped with a variety of very neat handles. In the north the spiral and meander are the leading motives and polychromy is freely employed; in Thessaly the patterns are purely rectilinear and are executed in only one colour. There are indeed a few features besides the fact of painting common to both areas: female figurines of clay were manufactured on both sides of the Balkan range, some of the celt types are similar, sun-dried brick [2] may possibly have been used in both provinces for building, and a stray stone seal from Thessaly might be compared to clay stamps from Transylvania and Bulgaria. But the types of figurine are far from identical, and the long porched houses of Transylvania do not yet appear in Greece.

If significant links between either European group and Asia be sought, the investigator is in the same quandary. The oriental material falls into a multiplicity of distinct cultures. Anau I constitutes a group apart no more closely related to Susa than to predynastic Egypt.[3] Susa I with the early painted pottery from Bushire on the Persian Gulf and from Southern Babylonia (Ur) forms

[1] See my *Dawn*, pp. 65–71 and 152–168.

[2] At Orchomenos I in Central Greece (Bulle, *Orchomenos*, pp. 19–20). In South Russia I now think that the inexplicable structures called *ploshchadki* may have been built of such brick.

[3] Frankfort, p. 76. If this author over-estimates the differences, it is certain that Myres, Moret, Langdon, and Pumpelly have greatly exaggerated the resemblances between Susa and Anau.

another distinct group to which Baluchistan and India may perhaps be added.[1] Susa II on the other hand is connected by its pottery with Northern Mesopotamia and even Palestine-Syria, but diverges fundamentally from its predecessor Susa I.[2] (Compare Figs. 14 and 15.) The position of the Chinese wares is still very uncertain. The two European groups show points of contact with all these distinct Asiatic families, but with none more than another.

At Anau the first settlers did live in mud-brick huts as did some European villagers and those of Susa and India, and, like the Thessalians and also some early Palestinian peoples and the Aegeans of Melos, buried children in jars under the houses. But they manufactured no figurines, used no seals or clay stamps, knew not the spiral motive and employed a ceramic technique and a set of vase forms very different from the Thessalian or North Balkan-Ukranian. At the same time they and all the other Asiatic vase-painters save those of China made use of perforated pear-shaped or spheroid mace-heads of stone which were unknown in South-east Europe at the period which concerns us. On the other hand if we come down to the third settlement at Anau (there were four in the oasis) the European parallels are more numerous; for both female figurines and clay or stone seals are encountered. But by this epoch the other links which were uniting Europe and Trans-caspia have dissolved; for the camel had by now been domesticated (he appears even in the second village), and painted pottery is rare, while monochrome vases made on the potters' wheel predominate.

But the figurines and seals and clay stamps appear at Anau in conjunction with other phenomena which are relatively southern. The south is evidently the home of a crescent-shaped copper sickle with a looped tang like Fig. 16, 1, with parallels at Kish in Babylonia and in Elam in period II and of the art of alloying copper with lead, which was Sumerian. Probably the use of the potters' wheel was learned from the same quarter. But some of the finds point to influence from the south-west. That is undeniably the case with a pin terminating in a double spiral—a Trojan-Cycladic type—and a beak-spouted jug which, if not inspired from Crete, would at least be Anatolian. The stamps and figurines may then have come to Anau from the same quarter and have reached Europe independently. As a matter of fact the clay stamps do

[1] The sherds brought from Seistan by Sir Aurel Stein are certainly very like the Babylonian and the site lies on the way from Mesopotamia to India.

[2] Ibid., pp. 43 f. M. Pottier takes the opposite view.

recur not only in Cappadocia,[1] but also in Troy where the pottery was not painted! Hence an immigration from Anau in its third phase is not requisite to explain the European analogies and is indeed extremely unlikely both on palaeontological and chronological grounds. If any migration connects Anau with the west, it is most likely to have been in the opposite direction; for this settlement is generally assigned there to the second millennium B.C., and we find in its ruins a hollow hemispherical button with a loop on the inside,[2] an ancient European type which we shall shortly meet in the Caucasus along with other undeniably European objects (page 124).

Further isolated parallels to the European material may be cited from other groups of Asiatic cultures, but always dispersed. Spirals are to be met in Armenia, the Punjab, at an uncertain date in Transcaspia and in the Honan province of China (Fig. 13), but in no case does a running spiral constitute the very basis of the ornament as in Transylvania and the Ukraine. The meander, equally common with the spiral in the latter regions, has so far only been reported from Kansu in China.[3] On the other hand naturalistic motives characterize the Asiatic pottery from Elam (Fig. 15), Syria, and Kansu,[4] but in Europe only appear as stray intruders in the geometric framework in the second cultural phase of the Ukraine. Conversely, the tall vase-supports and pedestalled bowls which are leading forms in the oldest North Balkan painted ware have convincing parallels in Mesopotamia, but among cult-objects which may be Sumerian and unconnected with the painted fabrics, and recur unpainted at Troy. Again, theriomorphic vases, found with painted ware in Europe, recur at Susa and the Punjab, but seem most at home in eastern Asia Minor and the Caucasus.

In conclusion, let us mention some unexplained parallels between China and South-east Europe. Some tripod vases from South Russia (possibly no older than period II) are exactly like those from all the Chinese sites [5] but find no analogies in the intermediate stations. Again the prehistoric villagers of China wore rings of mussel-shell, and similar ornaments are found with painted pottery in Baluchistan and Thessaly, while the prehistoric

[1] Chantre, *Miss. en Cappadoce*, pl. vi, 15; note the spouted vase; ibid., pl. viii; cf. also Frankfort, pp. 81 f.

[2] Pumpelly, vol. i, fig. 259.

[3] *Ymer*, 1924, loc. cit., fig. 1.

[4] Ibid., figs. 8–9.

[5] *M.A.G.W.*, liv, p. 73, fig. 12.

stone bracelets of China[1] have parallels in Egypt, Thessaly and Italy.

Enough has now been said to demonstrate that the attempt to crystallize out of the general complex in which painted pottery occurs, a single and peculiar group common to Europe and Asia leads to a cul-de-sac. We may still believe that this ceramic art was introduced into Europe from Asia, and that perhaps with more confidence than before, but we cannot isolate any specifically Eurasiatic culture associated therewith to contrast to an Eurafrican or Africo-Asiatic. So we come back to the abstract unity with which we started and to the same hypothetical wanderers as its vehicles. Now in no case can the earlier descendants, ethnic or cultural, of these assumed migrants be convincingly and unambiguously connected with Aryans. On the other hand their heirs can in several instances be shown to have entered into the composition of non-Indo-European peoples.

That is obviously the case with the predynastic Egyptians who, although surviving into historical times, left no traces of Indo-European speech in the Egyptian language. Of the earliest vase-painters of Susa I and South Mesopotamia it is not possible to speak with the same confidence. M. Pottier would indeed see in the former proto-Elamites, while Dr. Hall still thinks that the earliest prehistoric people of Ur may have been Sumerians.[2] Mr. Frankfort combats both these assertions, and his arguments are very powerful.[3] Even less can an ethnic label be attached to the inhabitants of Anau I.[3] But the people of Susa II are connected by a variety of traits[4] with an ancient population which has left its mark in the pre-Sumerian levels of Assur, in Cappadocia and North Syria, and whose artistic style survived in Palestine into Professor MacAlister's " Second Semitic " period[5] as M. Vincent[6] has recently demonstrated. That is to say this culture and this painted ware belonged to the population of the mountain zone encircling the Fertile Crescent on the North. We need not here ask whether that population should be designated " Semitic " or whether it did not rather represent Asianic " proto-Hittites "—some of its members

[1] On these see Arne, op. cit., and Andersson in *Pal. Sin.*, D, I, 1, p. 14; note also the celts there figured and the stone bead of fig. 4 which is rather like one from Anau III (or IV), Pumpelly, fig. 338.

[2] *Man*, xxv, 1.

[3] Cf. Langdon, *C.A.H.*, i, p. 362, for an opposite view.

[4] Frankfort, op. cit., pp. 62 and 70 ff.

[5] *Excavations at Gezer*, level III.

[6] *Syria*, v, pp. 91 ff.

wore a sort of pig-tail,[1] a style of headdress we have learned to know among the Hittites (page 28). It is enough for us that the cuneiform texts know no Aryans in these regions during the period of the early painted pottery [2] or, in the case of Palestine, only as isolated intruders in an essentially Semitic region. As for the Chinese pottery, the skulls found with it look quite like "the present inhabitants of North China" to Mr. Dudley Buxton,[3] and the tripod vases seem to be prototypes of the Li-tripods of bronze used under the early Chinese dynasties.

If so many of the earliest vase-painting peoples were not Aryan, it is highly unlikely that the initial masters in the art were such or that the migrants to Europe belonged to that stock. This conclusion is fortified by cultural considerations. The numerous female figurines in South-east Europe point to the cult of a Mother Goddess of whom Indo-European religious terminology preserves no reminiscence (page 81). In the economy of the vase-painters agriculture was fundamental—it has even been questioned whether the Susians of period I had any domestic animals at all—with the Aryans we suspect that it was only a secondary source of nourishment. When vase-painters had domestic animals, the pig was always prominent, sometimes the most prominent; [4] we have had to query the domestication of swine among the Aryans (page 83).

The idea that the diffusion of painted pottery in Eurasia was the work of Aryans remains a frankly attractive hypothesis. Some day it will be refuted or verified by further excavation in Iran and Central Asia. Till that happens the weight of evidence is against it, and we shall pursue our quest for some group of remains which can with greater confidence be connected with Aryan peoples.

But though the claim of these Asiatic immigrants to the name Aryan be provisionally rejected, their rôle in the formation of civilization in Europe needs a word of appreciation. Firstly must two waves of land-seekers be postulated? That would seem to be

[1] Andrae, *Die archaischen Ischtartempel in Assur*, pls. xliii and xlvii *c–f*; cf. Frankfort, p. 88.

[2] Cf. p. 23 above; the painted pottery of Cappadocia was very likely contemporary with the Semitic settlement there known to us from the Cappadocian Tablets. Dr. Christian indeed regards the naturalism of the Palestinian and second Susian pottery as a proof of Aryan influence, but almost in the same breath attributes the same quality in the metal work from A-anni-padda's temple at Tell el'Obeid to Semitic inspiration (*M.A.G.W.*, lv, pp. 190 and 193)! Both contentions are equally perverse.

[3] *Man*, xxv, 10.

[4] So in China, at Anau, in South Russia, Thessaly, and South Italy.

implied in the contrast between the first neolithic culture of Thessaly and the North Balkan. Yet the first band have left no traces of their passage either on the western coasts of Anatolia or in South Russia. It is still just possible that the two contrasted cultures belong to different branches of the same tribe, the peculiarities of the more northerly being due to intermingling with another ethnic group which we shall learn to know as Danubian. The colonists who established themselves in the Ukraine, Transylvania and Bulgaria must in any case be supposed to have come thither by land from Central Asia. That does not necessarily mean that they travelled along the steppe north of the Caspian and the Caucasus. There is another route from Central Asia south of the range through the valleys of the Kura and the Rion (the ancient Phasis) [1] and then along the Black Sea coasts. As a matter of fact painted pottery is said to have been found in the Crimea, and in the Araxes valley vases have been discovered [2] which, more than any others known to the author, resemble at once the European and Asiatic styles, although they apparently belong to a relatively late epoch. A journey through these valleys would help to explain the special analogies between the culture brought to Europe and that flourishing in the North Mesopotamian region. Whatever route they followed, the immigrants only began to settle down when they reached the extremely fertile loess lands, now the "black-earth" belt, on the edge of the forests on the western margin of the steppe (upon which ranged other more nomadic peoples).

Here in the valleys of the Dniepr, Bug, Dniestr, Pruth, Sereth and their tributaries they established their villages, tilling the marvellously fruitful soil and very likely adding to such head of domestic stock as they had brought with them by interbreeding with local species such as the wild swine. And very early indeed they crossed the Carpathians to settle on the head-waters of the Alt in Transylvania. And thus they introduced the "neolithic" civilization into Central Europe. It is, however, unlikely that they found these regions absolutely deserted and there are indications of early admixture with other races. The brachycephals, attested by the skulls from Bulgaria and Roumania mentioned above, may have, it is true, been numbered among the original migrants

[1] Cf. Casson in *B.S.A.*, xxiii, pp. 112 ff. His map shows how the Transcaucasian mounds lie along a line adjoining Anau and the black-earth belt of Russia.

[2] At Kizil Vank, *Izvestia. Imp. Arch. Komm.*, xxix (1909), pp. 1 ff. The pot from near Erivan, figured by Frankfort (pl. v, 1) and compared by him to fabrics from Susa I (!), is obviously allied to this group.

from Asia. But certain features in the material from the Ukraine, Transylvania and Bulgaria, most notably the spiral ornament on the vases, lead us to think that the Asiats were there amalgamated with other tribes of Mediterranean affinities and more lowly culture [1] with whom we shall soon become better acquainted. At a still later date signs of Nordic influence will be noticed among the vase-painters. Finally in Transylvania the peasants found themselves in a land of gold, and perhaps the command of this wealth brought them into commercial relations with the Aegean, Anatolia, and even Mesopotamia and Egypt; certain it is that civilization upon the banks of the Alt early blossomed forth into urban luxury.

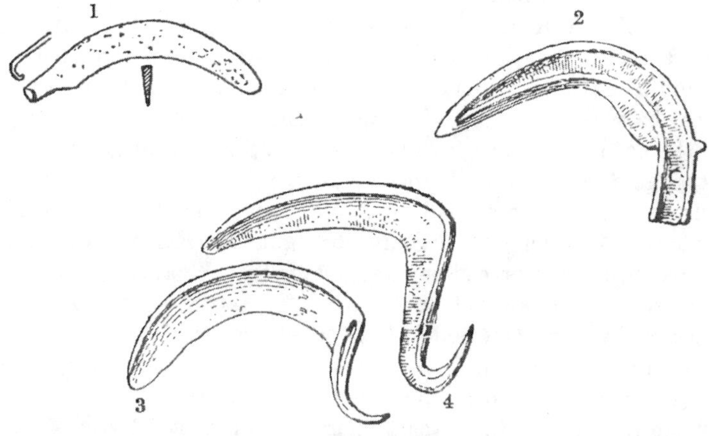

FIG. 16. Types of Sickle. 1, Mesopotamian (Troy VI); 2, European (Switzerland); 3, Transylvanian; 4, Caucasian.

Nevertheless the culture of the vase-painters in this area remained essentially Asiatic; even in the Bronze Age a sickle was there in use which diverged from all European models and was derived from the crescent-shaped type with looped handle that we have met in Elam, Turkestan, and Babylonia (Fig. 16). Perhaps it was owing to their orientalism that the remarkable civilizations of South-east Europe were eventually submerged by more truly occidental cultures.

[1] See the reservations made in *Dawn*, pp. 158–60; cf. Myres in *C.A.H.*, i, pp. 80 f.; he, however, is mistaken in imagining that the spiral was unknown to the first period. It is just in that period that this motive is found; in the later phase it is dissolved into circles and arcs.

3. *The Caucasus and the Iron Age in Europe*

There is yet another phase of cultural development in which the work of immigrants from Central Asia is in the eyes of some authorities discernible—that is with the inauguration of iron-working. Some would ascribe the introduction of the new metal to the Aryans as such, others would see in its bringers the last wave of Aryan invaders from Asia; the late M. de Morgan called them

1.

2.

FIG. 17. Engraved Bronzes of the Early Iron Age. 1, Girdle plate from Transcaucasia; 2, Plate of a Greek fibula.

frankly Celts. Could it be proved that the European Iron Age was in reality ushered in by an Aryan migration from Central Asia, even by the last wave of that migration, the general thesis of a Central Asian origin would be established; for after all it is only in the Iron Age that the Aryan peoples of Europe—Hellenes, Romans, Celts—are recognizable with absolute certainty.

It is to-day generally accepted that the earliest centres of iron-working on a large scale lay somewhere in the Hittite realm of northern Asia Minor (page 29). At the same time the Early Iron Age civilization of Central Europe, the so-called Hallstatt culture, shows such close affinities with that of the Caucasus that only a racial drift from one end to the other of the Danubian-South Russian plain will explain them. Now this Hallstatt civilization belonged to and was diffused by the Celts and Illyrians. Moreover, the Early Iron Age geometric culture of Greece and the Villanova culture of North Italy are, as we have already seen, connected by some authorities very closely with that of Hallstatt and attributed to the Hellenes and Italici respectively. If then the connections between the Caucasus and Central Europe do betoken a dependence of the latter area on the former the orientalist case will be well nigh established.

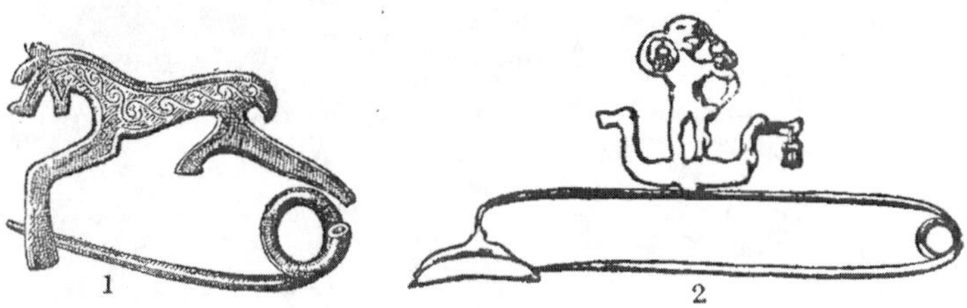

FIG. 18. Early Iron Age Brooches. 1, Koban; 2, Suessuola, Italy.

The parallels between the material exhumed from graves in the great cemetery explored by Bayern, Virchow, and Chantre at Koban [1] on the road across the Caucasus from Vladhivkaz to Tiphlis, and that from the necropolis of Hallstatt in Upper Austria and other sites in Central Europe, are indeed numerous and exact.[2] But they fall into two perfectly distinct groups. We have in the Caucasus on the one hand objects which recur to the West at Hallstatt or other contemporary cemeteries of the Early Iron Age and on the other types which in Europe belong to a distinctly earlier period, the Middle or even the Early Bronze Age. Such duality is scarcely compatible with the doctrine of a one-sided dependence of the West on the East.

[1] Published by Chantre, *Recherches anthrop. dans le Caucase*, 1885–7; cf. J. de Morgan, *Mission au Caucase*, 1889.

[2] An exhaustive list of these is given by Wilke, *Z.f.E.*, xxxvi, pp. 40 ff., but without taking into account chronological differences.

Let us consider first the former group. Both the folk buried at Koban and those interred or, earlier, inurned at Hallstatt were great horsemen, and there are many remarkable similarities in the bits and horse-trappings found at both cemeteries. Particularly striking are the openwork pendants often shaped like bells (Chantre, pl. xxvii, 9). The warriors of Koban and Hallstatt both used peculiar bronze or iron swords distinguished by a hilt terminating in crescent-like projections turned away from the blade (ib., pls. v bis, 2, vii, 2. Cf. our Fig. 25, 6).

Archaeologists call these weapons "antennae swords". In individual cases the sword-hilts from both areas were decorated with patterns formed by bosses in relief (ib., pl. v b). But the most distinctive common trait is to be found in the sphere of art; in

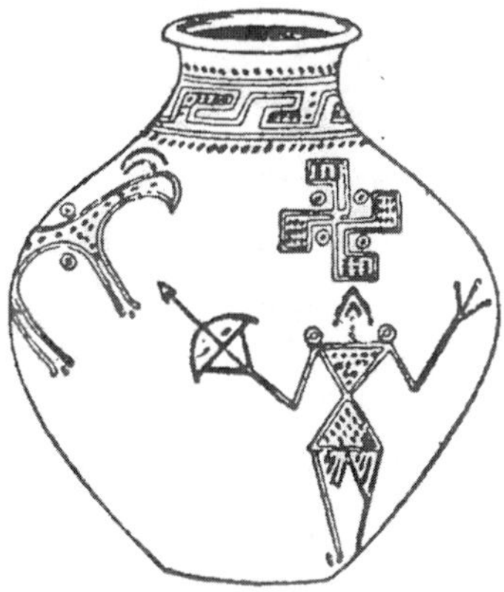

FIG. 19. Early Iron Age Vase. Hellenendorf, Transcaucasia.

both regions a striking decorative style characterizes the pottery and metal work. Bizarre animals—horses, dogs or even men—appear cast in bronze as pendants, or ornaments for chariot poles, engraved on bronze plaques and girdles and incised or painted on vases (Figs. 19–21). Even brooches (fibulae) are made with a dog's or horse's body (Fig. 18). At the same time this art was everywhere inspired with a veritable *horror vacui*, and the spaces between the naturalistic representations on plaques and vases are filled up with swastikas, meanders, spirals and concentric circles. It would be easy to amplify the list of analogies, but enough has been said to demonstrate the close connection of the two cultures. Add to all this that the Koban civilization is attached by other traits—the glass beads, the open-work technique on the bronze pendants, the style of inlaying sword-hilts and girdle-plates with coloured enamels and certain dagger types to the South Caucasian and Hittite civilizations, among which iron industry probably originated, and that the animal style itself is a survival of an older

FIG. 20. Early Iron Age Vase, Langenlebarn, Lower Austria. (Hallstatt style.)

local tradition (cf. page 184 below), and the idea that the Hallstatt culture is a transplantation to Europe of the Caucasian seems irresistible.

Yet the full facts do not justify such a conclusion. In the first place the objects from the Koban only explain a fraction of the Iron

Age material of Central Europe, let alone Greece and Italy. For instance, at Hallstatt the typical weapon was a very long sword with the blade designed for slashing, and in Europe this type goes back to the Middle Bronze Age (XIV–XIIIth centuries). In the Caucasus the swords are normally short—60 cm. is an exceptional length—and are usually designed for thrusting. Again in Central Europe the commonest form of axe-head was the "socketed celt", which was fitted on to the bent fork of a stick. This type is missing in the Caucasus, where the genuine axe-head perforated with a hole parallel to the shaft was in use. Again

Fig. 21. Early Iron Age Vase, Greece. (Dipylon style.)

the Hallstatt brooches belong to a different series, or an earlier stage of the same series, to those found at Koban.

Secondly some of the phenomena on which we have relied appear in Europe—not indeed on the Danube, but much nearer it than the Caucasus, in Greece—in an earlier context than at Koban. Thus iron was coming into use there in the latest Mycenaean age in association with the simple violin-bow fibula (Fig. 8, 2–3); in the Koban graves the earliest type of fibula is the derivative arc-shaped variety (Fig. 8, 4). The same remark applies to certain decorative motives and the list might easily be extended.

But thirdly the Koban culture is to a much larger degree dependent on the European Bronze Age civilization than the western Iron Age can be supposed to be dependent upon it. That is to say the Koban presents fully formed a whole series of types the evolution of which can be traced in Europe and nowhere else. For instance, one set of Koban daggers with a bronze hilt cast in one piece with the blade so as to leave a semicircular indent at the join are evidently imitations of the "Italian" dagger (Fig. 12, 4) which was diffused throughout Central Europe and as far east as Lithuania by the Early Bronze Age (1700–1400 B.C.). Most striking is a bronze sword of this pattern found at Mouçi Yeri just south of the range from the Koban, the hilt of which was composed of alternate rings of bronze and bone (Fig. 25, 7), for the same type is found in Denmark by the Middle Bronze Age (1300 B.C.). Again a whole series of the ornaments from the Koban tombs—notably the penanular bracelets with recoiled ends or terminating in opposing spirals or double spirals, wide arm-bands of bronze with four or five horizontal ridges on the outside and cylinder-shaped coils of wire terminating in spirals—belong in Hungary, Silesia and Denmark to the Middle Bronze Age, while some go back to the Early Bronze Age. So again the pins from the Koban in which the shaft expands above to form a wide flat head, raquet-pins as they have been aptly termed, are only a specialized variant of a type known already in Hungary at the end of the Early Bronze Age (before 1400 B.C.).

All these types and many others appear in the Caucasus fully formed, whereas in Central Europe the several stages of their evolution can be traced in detail. At the same time the Caucasian specimens are dated relatively to the western by the associated fibulae. The simplest of these have semicircular arched bows (like Fig. 8, 4). This is a secondary type in Europe proper to the Late Bronze Age or in Greece to the end of the Achaean period, and is therefore dated not earlier than 1200 B.C. The older type shaped like a violin bow (Fig. 8, 1–3) and belonging to the Middle Bronze Age in Europe and the Mycenaean period in Greece (1300 B.C.) is not represented at Koban. Hence the second class of Caucasian-Danubian parallels is constituted by types the appearance of which is earlier in Central Europe than at Koban. That is to say they travelled thither from the West, not *vice versa*. And lest there should be any doubt of the point a little amber (presumably of Baltic origin) has been found in the Koban tombs.

But this is not all. South of the range another series of necropoles have been explored, some of which represent the Koban culture in a later phase of evolution.[1] For instance, they contain fibulae which are an elaboration of Koban types, sometimes giving rise to a peculiar local variant in which the pin is a separate member and pushed through two catches in the bow. In these more southerly sites the influence of Central Europe is less apparent, but it is still occasionally noticeable. For instance, it is from this region that the "Danish" sword comes and a clay jug with excised ornament is remarkably like some Late Bronze Age vessels from Bavaria. All this goes to show that certain elements in the Koban culture came thither from the north-west and subsequently advanced further in the same direction to Transcaucasia.

There is therefore no longer any reason left for bringing the Celts or any other wave of Aryans from the Caucasus to Central Europe. However, the ghost of this idea still haunts Mr. Peake, and needs to be banished. This author, who agrees that the Koban culture was largely inspired by people come from Central Europe whom he accepts as Aryans, nevertheless holds that some of them returned from the Caucasus and brought with them the Hallstatt culture to the Danube valley.[2] He naively imagines that the visitants to the Koban, delighted with the iron that had been shown to them by natives living in the Transcaucasian valleys, galloped back across the 1,800 miles of steppe to exhibit it to their "relatives" in Hungary as a child might show a new toy to its mother. This is a hard saying. Historically, peoples like the Goths who reached the Caucasus from the West seem to have stayed there. And our Koban folk apparently crossed the range.

It remains to ask whether the first group of parallels—those between Koban and Hallstatt—really presuppose any direct connection. Now I do not think that there can be any question of direct Hallstatt influence in the Koban. All the most European types in the latter region are anterior to the Iron Age and some of them are missing from Iron Age graves in the West. On the other hand, as soon as it is recognized that the Koban culture has roots in the Central European Bronze Age civilization, it becomes clear that many objects common to Koban and Hallstatt are just survivals from this older period of community. For instance,

[1] See de Morgan, *Miss. au Caucase* and *Prehistoric Man*, fig. 66.
[2] *Bronze Age*, pp. 121 f.

curious adzes with lateral lugs found in both cemeteries are known in a rudimentary form in the Early Bronze Age of the Saale Valley and in the Copper Age of Italy. Again the hollow hemispherical bronze buttons with a loop on the inside, worn both by the Hallstatt and Koban folk, are met in Hungary in deposits assigned by Baron von Miské to the Early Bronze Age. Thus the similarities are largely explicable as parallel developments of a common cultural substratum in both areas.

The rest can partly be explained as the result of the influences at work in both regions but emanating from a common centre. The use of iron doubtless came to the Koban across the range from Asia Minor. There is every reason to suppose that it reached Europe from the same quarter. We have already suggested that it was transmitted to Greece from Anatolia. And since the amber trade was still in full swing, the knowledge of the new metal may well have been diffused up the Adriatic and into Central Europe as a result of that commercial intercourse. It is precisely on the amber trade routes that the earliest centres of the iron industry in Europe arose. But if that be so, other common phenomena—the open-work metal decoration for instance—may well have reached Hallstatt and Koban independently from a common centre in Asia Minor or Assyria. The influence of these regions is observable in the Early Iron Age both of Greece and Italy, and the most competent authorities [1] hold the Hallstatt culture to be posterior not prior to the earliest Geometric Age of Greece or the first phase of the Villanova culture in Italy. The typical Hallstatt bird meets us on a cup from the Tiryns hoard (page 52); perhaps this marks a resting-place in its flight from its oriental nest to Central Europe.

So the diffusion of iron working in Central Europe, like that of metallurgy in general, would be due to the fertilizing inspiration of commerce. The only ethnic movement which the parallels between the Caucasus and the Danube Valley presuppose is one from the West. And even so it must be borne in mind that South Russia was not an uninhabited desert. We shall later see that it was occupied by a mobile population well adapted to act as mediators in the transmission of culture.[2]

[1] Hoernes, *Urgeschichte der bildenden Kunst*, p. 436.

[2] A common ancestry in the latest painted ware of the Ukraine may ultimately explain the ceramic parallels between Hallstatt and the Caucasus.

4. *The Possibilities of the Anatolian Plateau.*

Archaeological evidence then fails to provide the expected support for the doctrine of a Central Asian cradle. But there is another corner of Asia which has put in a claim to be both the reservoir which supplied part of the neolithic population of Europe and the primitive habitat of the Indo-Europeans. The tendency among anthropologists in this country has recently been to locate the area of characterization of the brachycephalic Alpine race in the tablelands of Asia Minor.[1] At the same time the discovery of both *satem* and *centum* Indo-European languages on the fringe of the Anatolian plateau has induced Professor Sayce [2] to propose the transfer of the Aryan cradle from Central Asia to Asia Minor.

Now several migrations from that quarter into Central Europe are supposed to be detectable. According to Professor Myres [3] the first intruders into the uplands of Europe, which are as it were an extension of those of Asia Minor, brought with them the rudiments of agriculture and the habit of building pile-dwellings on the shores of lakes and swamps. As is well known, such pile villages are the characteristic features of the New Stone Age in Switzerland and Bavaria, and Professor Myres can point to survivals of the same style of habitation in Macedonia and the Caucasus in historic times. Moreover, the Alpine lake-dwellers were brachycephals. However, I cannot agree that the idea of constructing pile-dwellings was necessarily an importation from Asia nor that it gives evidence of an immigration from Asia at least in the period which concerns us.

Peculiar pressure from the environment must have been needed to impose upon primitive man the laborious task of erecting pile-structures to inhabit. Now the requisite conditions are fulfilled in North Europe after the glaciers had at length retreated; for they left a world of swamps and damp forests behind them which would almost force its denizens to construct some artificial resting-place. And as a matter of fact we find that the very early settlers on what was to be the Baltic, but was then a shallow mere, men who had not yet reached a neolithic stage of culture, did devise a sort of habitation from which the true pile-dwelling might have been evolved.[4] To find a dry place to lie down in and to be near

[1] Haddon, *Races*, pp. 26, 57; Myres, *C.A.H.*, i, p. 62.
[2] *Ramsay Studies*, p. 393.
[3] Loc. cit., pp. 72–5. Myres is mistaken in thinking that the earliest lake-dwellers had no domestic animals; the domestic species are best represented in the very oldest Swiss settlements. *Dawn*, p. 246.
[4] For full details see *Dawn*, pp. 212 and 245.

the fish on which they largely depended for nourishment, these pre-neolithic Baltic folk sometimes made rafts of logs and saplings on which they lived. And so did the direct descendants of the same people in Sweden and Denmark in full neolithic times. But the latter had made improvements. The raft of logs was first converted into a fixed pontoon by posts at the corners. But such a pontoon soon became waterlogged, and fresh layers of logs had repeatedly to be added to form a dry floor, till at length a piled stack resting on the lake bottom was created. Such primitive structures are known both in Denmark and in Switzerland and Wurtemburg. In founding new settlements some genius hit upon a modification which considerably economized labour. Instead of making the foundations of your village out of a stack of many horizontal logs, you laid down a single platform resting upon rafters supported by upright piles, and this required far fewer trees laboriously felled with stone axes. So perhaps arose the classical pile-dwelling.

Now the pre-neolithic raft-builders of the Baltic had included brachycephals like the neolithic pile-dwellers of the Alps. At the same time there is reason to believe that a kindred stock of hunters and fishers was widely diffused throughout the forests and marshes of northern Europe in early post-glacial times. Some of these we may suppose retreated, perhaps up the Rhine, to the highland zone in pursuit of fresh-water fishing and such game as the chamois at the time when the salt waters of the North Sea made their way into the Baltic depression and the climate became milder. And in the uplands they found fresh-water lakes on which they settled, developing their domestic architecture through similar phases to those traceable among their kinsmen in Scandinavia. Thus the Swiss lake-dwellings are explicable without assuming any invasion from Asia Minor in neolithic times.

Moreover, there is positive evidence against the hypothesis of such a migration. In the first place the lake-dwellings of the southern parts of the Alpine zone, Carniola, Bosnia and Macedonia, seem all distinctly later than those of Switzerland, Wurtemburg and Scandinavia, not earlier as might be expected had the pile-dwellers come from the south-east. Secondly the neolithic elements in the Alpine and Swedish lake-dwellings are essentially different; the common features are only the architecture and certain "palaeolithic" survivals—bone harpoons, phallange whistles, and so on. That means that the neolithic arts had not been introduced with the custom of pile-dwelling, but had been acquired separately by the several groups of pile-dwellers from other more advanced peoples. Those of Scandinavia were in fact instructed by the megalith-builders of the coasts, those of Switzerland and Wurtemburg by the Danubians [1] whom we shall next consider. Thus we see the original Alpines as a food-gathering folk of the forest, swamp and mountain who only acquired the "neolithic" arts from more progressive neighbours. Hence, if they came from Asia Minor at all, they did not come as Aryans.

But those Danubians [2] whom we have just mentioned may themselves be immigrants from Asia Minor. They rival in antiquity the vase-painters from Asia and may have mingled with the latter in South-east Europe. It is certain that they spread from the Danube valley far and wide in Central Europe, diffusing the knowledge of domestic animals and cultivated plants to Little Poland, Silesia, Central Germany, the Rhineland and Belgium. Now some traits connect these Danubians with Asia Minor; in particular their clay vessels are evidently imitated from gourds. The gourd will not harden north of the Balkans, so that the prototypes of the Danubian pottery must be sought further south. Professor Myres [3] has adduced grounds for the belief that the primary focus is to be looked for somewhere in western Asia Minor or Syria, where gourd-like forms long persisted and in some cases are in use to-day. Of course, the Anatolian and Syrian fabrics are not identical with the Danubian and cannot be looked upon as prototypes of the latter, which exhibits peculiarities which might lead one to imagine a survival of palaeolithic art in the region. The most that can be admitted is a generic kinship with Asia Minor. And at the same time other Danubian peculiarities are distinctively Mediterranean.

To clarify the issue it is desirable to ask to what race the Danubians belonged. Mr. Harold Peake holds that they were Alpine brachycephals, and his view has been followed by Myres and Fleure. This supposed brachycephalism is adduced as additional evidence for the Anatolian origin of these people. But Mr. Peake's view seems to rest on a misconception; I can find no

[1] Myres, loc. cit., p. 75, agrees that the Alpines acquired the domestic stock from the Danubians. He does not hold that the original invaders had been fully neolithic, but seems to credit them with the rudiments of agriculture.

[2] On these see *Dawn*, pp. 171–6.

[3] Op. cit., pp. 77 f.

evidence for the presence of short-heads on the Central European loess lands at this epoch. The skeletal remains are indeed exiguous; still a few graves with contracted skeletons or stray skulls have been found in Serbia, Moravia and Lower Austria accompanied by Danubian pottery and artifacts. Not one of these skulls is brachycephalic; all are moderately long-headed and moreover agree in this and every other respect with a much larger series of skulls found with identical pottery in slightly later graves of Central and South-west Germany. It only remains therefore to ask to which dolichocephalic race these Danubians belonged. The late Dr. Schliz assigned them to the North European race, a branch of what became the Nordic stock. And no doubt true Nordics do appear mixed with Danubians, but only at a later date. The earliest skulls which concern us here belonged to short individuals, and resemble in several respects, as Schliz himself was forced to admit, Sergi's Mediterraneans more than the tall Nordics.

The correctness of the latter attribution is borne out by the markedly southern character of the Danubians' cultural heritage which links them with other Mediterranean stocks. Not only does their pottery imitate the southern gourd, but the black fabric and the incised decoration suggest vague comparisons with Cretan and North African wares as much as with Anatolian. Again the Danubians made female figurines of clay, and these show a tendency to steatopygy just as do early figurines in Crete and Egypt, and this feature is to-day counted a mark of beauty among the Bushmen of South Africa. Moreover this continental people even in the heart of Central Europe continued to deck themselves with the shells of a Mediterranean mussel—*Spondylus gaederopi*. Finally the one weapon found in the villages of the first Danubians is a mace-head formed of a flat stone disc perforated at the centre and generally sharpened at the edges. In contrast to the piriform or spheroid types this is a rare form of mace, and originated somewhere in the immediate vicinity of the Nile valley, since the type was current in Egypt in predynastic and the earliest dynastic times, but perhaps nowhere else in the Ancient East.

We must then regard the Danubians as a branch of the Eurafrican race. That does not exclude the possibility that they came to Europe immediately by way of Asia Minor, either crossing over by the Dardanelles and Bosphorus or even travelling at a still earlier date by the old land-bridge where the Archipelago now lies. Some such hypothesis would explain the ceramic similarities between Danubian and Anatolian pottery—similarities which extend also to Crete and the Cyclades—and the recurrence of certain dolichocephalic types on both sides of the Aegean and north of the Balkans, which Serbian students have designated by the doubtful name of Pelasgian. Plainly the migration in question must have been very early—anterior even to the advent of the first band of Asiatic vase-painters. One really wonders how much of the "neolithic" civilization these proto-Danubians brought with them ready made. All that is proved is the use of gourds as vessels, the tradition of a Mother Goddess, not necessarily represented in idols of baked clay rather than, say, wood, an affection for a particular shell and a very archaic type of weapon. Since the typical Danubian "celt" of polished stone is found at least as far south as Thessaly and was in reality above all a gardening tool, a hoe, the cultivation of cereals in a garden plot may be added to the list. It remains possible that the Danubians' animals and their ceramic technique were borrowed from the Asiatics of Transylvania. At the same time the spiral ornament used by the latter was probably inspired by Danubian models which were not necessarily applied originally to clay vases. But here the possibility must be borne in mind that some survivors of palaeolithic tribes who, ages before, had decorated bone with spirals, still persisted in the plains of Central Europe and had mingled with the newcomers from the south.[1]

Should we then give the title Aryan to this Mediterranean stock as it appears in Anatolia and the Danube valley? There are no very cogent grounds for so doing. *Qua* Mediterraneans the Danubians were not Aryans. Nor do they constitute a substantial link between Asia Minor and Europe. In the former region they were at best but passers-by. And their primitive culture is too vague and inchoate to be called Aryan. All they brought with them were the rudiments of a cult, some simple implements and weapons and a few grains. Thus equipped the invaders of Central Europe created their own culture on the fertile loess lands of the Danube basin inspired perhaps by their Asiatic neighbours and that southern trade to which their superstitious attachment to a Mediterranean shell impelled them. Whether this people developed into Aryans there in Central Europe is a question for subsequent discussion.

[1] Cf. Menghin in Hoernes, op. cit., p. 774.

However, relations between Europe and north-west Asia Minor did not cease with the passage of the proto-Danubians. Out of the shadowy Anatolian culture, the relations of which to Central Europe have just been discussed, there arose by differentiation and concentration a more substantial civilization, best known by the remains from the second city of Troy (Hissarlik). This civilization undeniably influenced Thrace, Macedonia, the Danube valley, Thessaly and, through Thessaly, South Italy at a period subsequent to the descent of the second wave of vase-painting peoples into North Greece—i.e., between 2500 and 1800 B.C. The points of contact between Troy and Thrace have already been enumerated. In Thessaly the testimony to influence from Troy is in the first place a series of vases of which the most unambiguous are high-handled cups. In Central Europe these are found, not in the oldest Danubian graves, but in those of the second period from Hungary to Silesia and Bavaria and stretching well across the Illyrian mountains to Italy. They are certainly clay copies of metal vases of Troadic type. And with them are associated in Central Europe copper ornaments in the form of two spirals linked like a pair of spectacles,[1] another Trojan pattern. Other types of objects also found at Hissarlik, without being peculiarly Trojan—spheroid mace-heads, perforated stone axes and the most primitive copper axes—also make their appearance both in Central Europe and Thessaly about the same time. At a rather later date some current from the south-east carried up the Danube valley certain types of pin, earrings and the curious Cypriote dagger with a looped tang (Fig. 12, 2), all of which recur at Troy. Nor is this all, Troy II was in not altogether one-sided relations with the East. The pin with a double spiral head met at Anau III (page 111) recurs at Hissarlik and, if not specifically Trojan, was at least Aegean rather than Mesopotamian.

Finally on both sides of the Aegean, in each case about 1800 B.C., possibly allied phenomena appear, the seeming parallelism of which might be explained as the result of emanations from a single centre in Asia Minor—I refer to cist-graves of large stone slabs enclosing contracted skeletons, accompanied in each case by high pedestalled bowls, found among the houses of Carchemish on the Upper Euphrates and those of Orchomenos and other towns in Greece (cf. pages 27 and 60).

In all these directions we have points of similarity amidst

[1] *Dawn*, p. 179, fig. 79, 1.

differences. In Thrace the Troadic types are in a minority as compared to the peculiar local forms. In the Danube valley they appear alongside survivals of the earlier culture and contributions from other centres lying outside Asia Minor. It is just the same in Greece, and in particular the immediate derivation of the cist-grave culture from Troy is by no means certain; it is indeed quite as likely that the Minyan ware of Troy is due to influence from the north-west. Neither the cist-graves of Carchemish nor their contents can be regarded as immediately derived from Troy and the technique of the pottery found in them and the forms of the bronzes differ at once from the Trojan and the Greek. Finally the parallelism between Troy and Anau may be due to the influence of some intermediate culture upon both centres.

Nevertheless we have at this juncture a degree of cultural interconnection between wide areas in Europe and Hither Asia never hitherto nor subsequently attained. Furthermore the context in which the vestiges of this interconnection are detected is in some cases quite possibly Aryan. In the European cultural complex certain elements, which we have designated Troadic, persist and reappear in the Italian *terremare* which we have agreed to regard as the earliest monuments of Aryans in the Apennine peninsula. In Greece and Macedonia the culture in which our Trojan parallels appear is continuous with that which on one hypothesis might be assigned to the Hellenes (page 60). As we remarked above, traces of connexion between western Anatolia and Thrace, such as the traditions concerning the kinship of Phrygians and Thracians imply, are discernible at no other period till well on in the Iron Age, while certain types already current in north-western Asia Minor survive to emerge again in the barrows of Gordion which undoubtedly belonged to Phrygians. The Hittites had been exposed to Aryan influence some time in the second millennium B.C., and it is to Hittites that Mr. Woolley ascribes the cist-graves round Carchemish. Finally Indo-Iranians were wandering about into the north of Mesopotamia somewhere about this time and must later have embraced Anau in their domain.

Thus the links with western Anatolia which might be established through Troy would connect up a number of areas subsequently occupied by Indo-Europeans. It can, of course, hardly be contended that Troy was itself the centre of a proto-Aryan empire; it may very reasonably be demanded whether the Anatolian civilization of which Troy was one peripheral manifestation was not proto-

Aryan. To reach an answer to that question we are forced to rely very largely on the material gathered at Troy itself, supplemented by such inferences as phenomena observed at the opposite end of the plateau of Asia Minor enable us to draw. And it must be remembered that Troy is a mound of nine superimposed cities, the objects from each of which the excavator, Dr. Schliemann, did not very accurately distinguish.

In the civilization of Troy II, which is what here concerns us, a multiplicity of influences converge. We meet first types reminiscent of neolithic Crete, survivals of that earlier age to which reference has been made above, together with signs of the inspiration of the Bronze Age Minoan civilization. Then indebtedness to ancient Mesopotamia is attested in the use of brick for the fortifications and in the whole inventory of metal types. But the civilization of Troy in its more intimate aspects, its pottery for instance, is quite distinct from the Sumero-Akkadian and also from the intermediate culture of Cappadocia where the vases were painted. The ceramic evidence attaches the Troad to a more westerly culture, the roots of which are discoverable in Cyprus and North Syria—Myres' red-ware province. It can only be Anatolian culture in this latter sense that must concern us here.

But there are conclusive reasons for denying to it as a whole, just as much as to the Sumero-Akkadian or Cappadocian civilizations of the third millennium, the title Aryan. In the first place Anatolia was the very heart of the Great Mother's realm.[1] Was she not represented on the oldest cult monuments from one end of the plateau to the other, from Troy and Cyprus to Assur ? Was not her cult characteristic of the region at all epochs ? Did not even the Aryan Phrygians have to admit her to their pantheon ? It is inconceivable that any people coming from Asia Minor should have lost all recollection of her. Secondly in historical times not only was the eastern portion inhabited by non-Aryan peoples, but also on the western coasts of Anatolia dwelt remnants of cognate stocks, Leleges, Carians, Lydians, and so on.[2] At the same time the native topographical nomenclature of the whole region is non-Indo-European, but includes parallels, not only to Fick's pre-Hellenic names in Greece but also to those Asianic names occurring already in the third millennium B.C. on Cappadocian tablets. Hence it looks as if an Asianic population occupied

[1] Meyer, *Reich und Kultur*, p. 90.
[2] Ibid., p. 125; cf. Sundwall in *Klio*, 1911, pp. 464 ff.

the whole of the plateau. The Anatolian culture as a whole should be ascribed to this stock, and we may recognize their descendants among the Dardanian opponents of Rameses II, who resemble Hittites in some features (Plate VII, 1, cf. page 64).

But though this Anatolian culture cannot as a whole be regarded as Aryan, we were perhaps wrong in calling Troy a peripheral manifestation of it. The European parallels all refer to the north-west corner of Anatolia. Perhaps we should confine our attention to that region in seeking their roots. At least by the XIIIth century this area found formed a politically isolated unit contrasted to the rest of the land mass as we saw in Chapter III. But if we do look to the north-west corner of the Troad and its immediate hinterland, as opposed to the rest of the Anatolian promontory, a very surprising result awaits us.

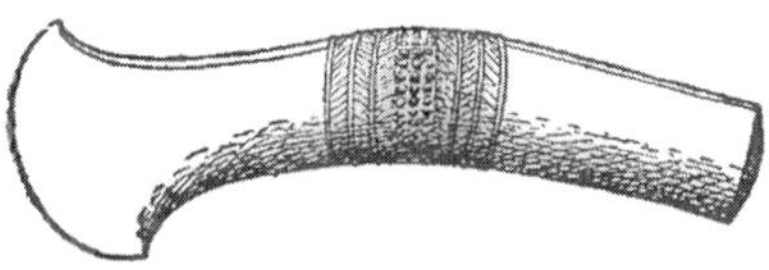

FIG. 22. Decorated Stone Battle-axe, Troy II.

When we make abstraction of the elements presumed to be common to the whole land mass and concentrate attention upon what is peculiar to its north-western corner, Troy no longer seems the Asiatic root of an European culture, but a branch of an European stem pushed across the Straits. The palace of Troy's kings was a *megaron*, a long narrow room with a central hearth and a pillared porch on the short side; the earliest dated examples of this sort of house come from Transylvania, and we meet the same type in Wurtemburg by 2000 B.C. (Pl. VIII, 1), and rather earlier in Thessaly. Again the Trojan kings wielded as symbols of their power heavy battle-axes of noble stones, superbly polished and richly carved (Fig. 22). Stone battle-axes are indeed very common throughout the ruins of Troy and recur at the contemporary cemetery of Yortan in Mysia. Such clumsy weapons are strange things to find in a Bronze Age town; in the rest of the Aegean area, in southern Asia Minor and in Mesopotamia, they are virtually unknown. But in Europe from the Volga to the Rhine they are scattered about in profusion and all the varied Trojan types are there represented. These European axes in Troy cannot (as I once thought) be explained by trade. Why should a people rich in metals import such barbaric weapons ? Why should they be symbols of kingly power ? Surely they are the monuments of an intrusion from Europe of a people accustomed in a wilder environment to swing such mighty hammers. And it is precisely this element which distinguishes the civilization of north-western Asia Minor from the general "Asianic" cultural background to which it was so deeply indebted. To the wielders of those ceremonial axes might perhaps be attributed the erection of that sovereign power which has transformed the village of Troy I into the city of Troy II and ultimately welded the heterogeneous tribes of the region into a compact confederacy. And so the Troad and its hinterland becomes part of the great European battle-axe province extending from the Baltic to the Black Sea. At the same time if the *Takrui* who attacked Egypt in 1192 B.C. be Teucrians from the Troad,[1] they attest the presence of men of European aspect side by side with the Armenoid Dardanians (Plate VII, 2).

So our question now assumes a new complexion. Is it with Troy as thus Europeanized or with the "Asianic" substratum that the Thessalian, Balkan, Italian and Danubian cultures are somehow allied ? On the one hand in Hungary the graves where signs of parallelism with Troy—high-handled cups, spheroid mace-heads and spectacle spirals—are first noted cover the remains of that same tall dolichocephalic race as wielded the battle-axes of Scandinavia, Germany and Russia (page 174), and that race was essentially European. On the other hand Anatolian culture as a whole did not penetrate into Europe. The ceramic parallels we have enumerated are in effect limited to imitations of Trojan metal vases. Such imitations, as well as the metal spirals, pins and earrings later found in Central Europe, may well be the result of trade. The Trojans were in possession of tin, since 10 per cent. of that metal entered into the composition of their bronzes ; they may well have been the inventors of this alloy, prompted by familiarity with the technique employed much earlier by the Sumerians of mixing copper with lead in the same proportions. The Trojans most probably imported this tin from Bohemia. That does not mean that they invaded Central Europe any more

[1] This is very doubtful, see p. 74 above.

1. DERDEN

2. TAKRUI

3. PULESATHA

[face p. 134

than amber in Minoan tombs and Minoan metal-work or clay imitations thereof in Thuringia imply a Minoan colony on the Elbe or than Greek vessels from Denmark denote a Hellenic colony on the North Sea. Individual pioneers from Troy may certainly have been the first to stumble upon the mineral wealth of Bohemian soil or that may have been the work of our Prospectors from the west (page 99), but the exploitation of the deposits was in the hands of the local population. In any case the Trojan pins and Cypriote daggers transmitted through Troy, which are found in the Danube valley, are landmarks on an ancient trade route. And we may believe that commerce flowed along that road till the fall of Troy II which seems to coincide with the diversion of Aegean-Bohemian trade to the route terminating at the head of the Adriatic somewhere between 1900 and 1600 B.C.[1] Trade rather than migration of peoples will then explain most cases of influence from Troy upon Southern and Central Europe.

But with these admissions the whole theory that the Thessalo-Illyro-Danubian culture was the reflex of an invasion from Asia Minor collapses. The cultural continuum is no longer just a projection of Asia into Europe ; it may equally well be designated an European culture with an Asiatic enclave : the Oriental connections are connections with the Troad as cut loose from the Anatolian cultural complex as a whole. The question of their final explanation is transferred at once to European soil. From that standpoint we must ask in the next chapter how far the unity which embraces North Greece, Macedonia, Upper Italy and the Danube valley is infused with the same elements as Europeanized Troy and so whether it is Aryan.

5. *The Claims of Asia Reviewed*

The conclusions of the foregoing paragraphs afford very slender support to the hypothesis of an Asiatic cradle for the Aryan people. The material available in Europe to the prehistorian does not disclose, as was once expected, wave upon wave of Asiatic immigrants bringing with them the civilization of the Ancient East. The neolithic population of Europe was very largely descended from the palaeolithic which already included both long and short-headed types. The brachycephalic invasion is receding into an

[1] *Archæologia*, lxxiv, p. 174.

ever remoter past. If hordes of Asiatics did drift westward during the geological present they have left singularly little evidence of their advent and so are unlike the later Scyths, Sarmatians and Mongols. In "neolithic" times only one tide of migration from Asia could be recognized by archaeological means. It brought the mysterious cultures with painted pottery to Thessaly, Transylvania, Bulgaria and the Ukraine. With the same movement were to be associated certain very important contributions to the civilization of South-east Europe, notably the introduction of agriculture and domestic animals of Asiatic species, oxen of the Urus breed and the turbary sheep. But the area occupied by these intruders was as restricted as that reached by the Asiatic invaders of the historical times; we found no reason to suppose that the propagation of the new arts was the work of the Asiatics as such. Save in a few corners like Bulgaria they seem to vanish while the new arts were taken over and elaborated by other tribes. Nor could we honestly identify the vase-painters with Aryans.

At the same time we have postulated an invasion of Central Europe by Mediterraneans, come perhaps through Anatolia, just as other streams of Eurafrican peoples were reaching Western Europe across the Iberian peninsula even in palaeolithic times. To the Mediterranean invaders we attributed the elaboration of the Danubian neolithic civilization probably assisted by culture contact with the Asiatics. But again we saw no reason to describe the proto-Danubians, *qua* strangers to Europe, by the epithet Aryan.

The third great impulse which affected Northern and Western Europe and some points on the Black Sea coast early in the New Stone Age, the megalithic culture, seemed even less to answer the requirements laid down for the Aryans. If and in so far as it was foreign, it was Eurafrican in character. And it was only in a minor degree the result of a racial drift. The idea of the megalithic tomb and the associated cult of the dead were very likely brought in the first instance by navigators from the southern shores of the Eastern Mediterranean imbued with Egyptian eschatology, if not themselves Egyptians. But a colonization of the coasts on a large scale by megalith-builders is not to be thought of; the actual settlers were few, but they instructed the natives in their religion and in some of the neolithic arts, notably the domestication of short-horned cattle. Neither the strangers who brought the cult of the dead and megalithic funerary architecture nor the Eurafrican aborigines who adopted and propagated them in Spain, France and Britain can on any grounds be regarded as Aryan. And with the dolmen-builders the list of Europe's invaders during the early neolithic period is closed.

In the sequel there is little room for immigrations on a large scale, though shifts in the population within Europe itself were frequent. And in this epoch, the full neolithic age of our continent, intrusions from Asia are not traceable with any certainty. In particular the one band of brachycephalic migrants whom we could detect came not from Asia but, immediately at least, from the Iberian peninsula, whence they brought the bell-beaker to Central Europe. Thereafter the development of the Bronze Age was self-contained and rapid till socn it was not Asiatic weapons and ornaments which were imported into Europe, but European types which migrated to and implanted themselves in the Near East as our excursion to the Caucasus showed.

But if our search for Aryans has so far been abortive, the results of the chapter were not merely negative. We beheld the founding of the new civilization in Europe, we witnessed the addition to the old palaeolithic stocks of new ingredients come from Asia and Africa and estimated the culture of the newcomers. It was in no case beyond that inferred for the Aryans. Hence if the extraneous elements did not themselves become Aryans in Europe, their culture was not such as to offer serious obstacles to Aryanization by conquest or absorption in the Aryan people.

CHAPTER VI

DID THE ARYANS ORIGINATE IN CENTRAL EUROPE

To-day the Asiatic hypothesis has been abandoned by most linguists; the last chapter showed that the archaeological evidence also led away from it. With the reservations made above no migration from Asia is discoverable which can with any probability be connected with Aryans. We are thus encouraged to follow the philologists on to European soil.

We have seen further how the neolithic population of Europe constituted a veritable mosaic of races. Culturally a still greater diversity reigned. From neolithic times the continent may be divided into a number of provinces each exhibiting its own material peculiarities. In which of the nascent civilizations of neolithic times shall we seek the first centre of Aryan activity? It makes no difference for us whether the authors of the several cultures were indigenous or intrusive. Aryanism grew up out of a racial mass, which must have been at some time not yet Aryan. But though the racial antecedents of the inhabitants of the several areas of neolithic Europe do not provide a criterion for excluding any of them from our survey, some regions may on other grounds be omitted. The consensus of opinion among historians and philologists allows the Mediterranean basin to be eliminated; that area, populated originally by various branches of the Eurafrican stock and owing the foundation of its civilizations to maritime commerce with the early cultural centres of that race, Egypt and Crete, was only Aryanized late in its prehistory. The same general agreement justifies us in passing over the great West European cultural province where the principal racial element was again Eurafrican, and the chief formative influence the Mediterranean megalith culture. In fact most investigators look to one or more of three regions, North Europe, East Europe, and Central Europe. We shall begin our survey with the last-named region, because the discussions of the preceding chapter were tending to conduct us up the Danube valley. We do so with all the more alacrity, since the claims of this region have been brilliantly championed by Dr. Giles in very recent times.

1. *Dr. Giles' Hypothesis and the Danubian Peasants*

As the area where the Aryan people were differentiated, Dr. Giles[1] has proposed the loess lands between the Carpathians on the east, the Balkan mountains on the south, the Alps and the Böhmer Wald on the west and the Erzgebirge and northern Carpathians on the north. Here he thinks the environmental conditions of the primitive Indo-European culture, in which, on his view, agriculture was just as important as stock-raising, are best satisfied and hence the Aryans, or *Wiros* as he prefers to call them, would have spread throughout Europe and to Asia. The exodus must he thinks have begun about 2500 B.C., and the route followed by the eastward migrants would be across the straits of the Bosphorus and Dardanelles and over the highlands of Asia Minor.

The distinguished philologist is not an archaeologist, and makes no attempt to trace his "Wiros" with the aid of material remains. But the area he has outlined, was in fact the centre of a distinctive culture, the development of which falls within the chronological limits he has laid down. This culture, which may be called Danubian or more precisely Danubian I, was the creation of those early Mediterranean colonists whose advent was discussed in the last chapter, and who, as we there saw, may early have been mixed with descendants of palaeolithic tribes and influenced from the east as well as the south. These people made their settlements exclusively on the loess, a very fertile soil that covers the plains of Central Europe to a considerable depth.

We may picture[2] these Danubians living as peasant cultivators in the fertile valleys. Their small unwalled villages were always planted in proximity to streams, and consisted of groups of half-subterranean huts (cf. p. 86). Near by were small garden plots, roughly cleared by stone axes and hacked up by stone hoes (the typically Danubian "shoe-last" celts) to receive the grain, which on the loess would flourish even under these summary methods. In the parklands of the adjacent slopes, grazed the peasants' herds—cattle, sheep and swine—watched by the village children,

[1] *Cambridge History of India*, i, pp. 68–70.
[2] For the evidence on which this picture is based see my *Dawn*, pp. 171–6.

just as is done in Galicia to-day. Perhaps the horse had been already domesticated[1] to aid in the pursuit of straying beasts. A little fishing gave variety to the villagers' diet, but the game from the primaeval forest, the haunt of bears and wolves, does not seem to have been hunted. The community was small with no regular division of labour. Within it the women doubtless

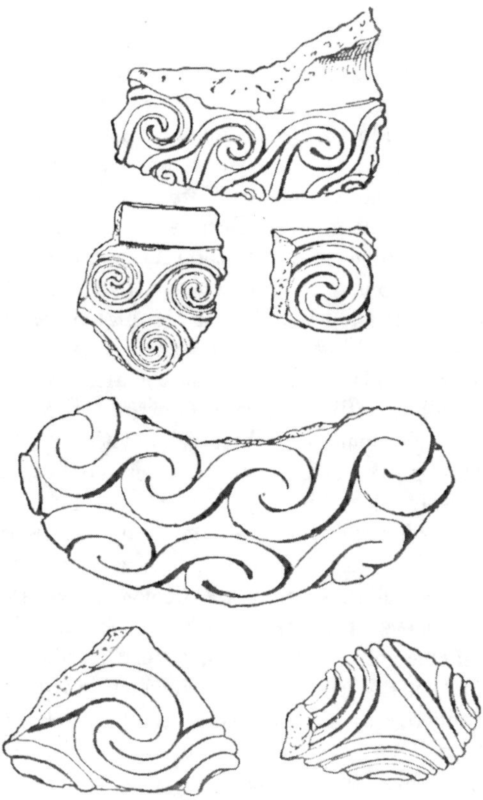

FIG. 23. Decoration of Danubian pottery from Butmir in Bosnia.

made the pots, imitating with feminine conservatism the gourd vessels of their remote ancestors. They decorated their clay vases tastefully with spirals and meanders (Fig. 23). The religion of the cultivators certainly included the worship of a Mother

[1] No positive evidence is, however, forthcoming for domestic horses in the Danube valley in period I. Horse bones found with Danubian I pottery in the Rhineland may be much later.

Goddess, clay models of whom adorned the huts. Beyond the garden plots would be other villages not very far away, and in the absence of weapons (save for the disc-shaped mace-heads) we may believe that the several groups lived in mutual amity. But in the background loomed the primaeval forest, almost impenetrable in summer and very perilous in winter.

Despite the forest, the Danubians spread far and wide. Even the fertility of the virgin loess might become temporarily exhausted under their rudimentary methods of cultivation. In any case the natural increase of the population under such favoured conditions, necessitated an even wider expansion of colonization. New generations went forth from the village, as in the Sacred Spring of the oldest Romans, to found fresh villages and bring fresh land under tillage. The colonists generally followed the river valleys and were aided in the transportation of their simple belongings by dug-out canoes. Ultimately, as in the course of centuries a gradual process of mild desiccation thinned out the woods, the Danubians crossed the Böhmer Wald and entered the Rhineland. But during all their history, peaceful communication was maintained throughout the whole area. The *Spondylus* shell bracelets, to which we have already referred, must have been handed on from village to village, perhaps in a ritual exchange of tokens like the *kula* traffic of the Pacific Islanders. As an incident in this commerce the Danubians became possessed of copper trinkets.

Moreover, the Danubians came in contact with other ethnic groups in the course of their wanderings. On the mountain slopes to the west, whither they had repaired in pursuit of the chamois and the deer, lived a sparse hunting population descended as we saw (p. 126) from pre-neolithic stocks. Here and there these wild hill folk borrowed from the Danubians elements of a higher civilization and gradually settled down. In Germany and Poland, the Danubians soon met tall men of the Nordic race, probably already pastoralists. In some cases the two stocks amalgamated and the Nordics assimilated the cultures of the peasants as on the Rhine; elsewhere the Nordics established themselves as overlords among the cultivators and eventually imposed upon the peaceful Danubian culture their own more warlike one. To the east the Asiatic vase-painters were established and interaction between them and the Danubians had begun very early.

Were the Danubians then Aryans, Dr. Giles' "Wiros"? First let us consider their relations to other areas later occupied by Aryans. In Europe a consideration of the question from this side leads to very satisfactory results. The Danubians occupied a considerable part of Poland, the whole of South Germany from the Oder to the Rhine and even pushed into Belgium, and possibly to East Prussia as well. Throughout this wide area the foundation of culture was Danubian even where racial intermingling took place. In the Alpine zone the lake-dwellers were deeply indebted to the Danubians for their arts and crafts; why not for their speech too? To the south Illyria was originally an appanage of the Danubian province as the finds from the celebrated station of Butmir in Bosnia show. And since the Italici of the *terremare* came from the Danubian side of the Alps (p. 71) a Danubian element may be assumed there too. Eastward many German investigators, such as Schliz, Hubert Schmidt, Menghin and Kossinna, would actually treat the Transylvanian and Ukrainian painted pottery as the work of Danubians and, though we cannot accept that thesis, we have admitted the possibility of a Danubian admixture among the vase-painters from Asia and therewith of a Danubian element in Thrace and Thessaly. To this extent the real or possible distribution of Danubians coincides quite well with that of the earliest Indo-European languages.

Relations with Asia are much more hard to find. The Danubians had a well-defined culture and art of their own. It should then be possible to point to some monuments of the various waves of migrants across Anatolia, invoked by Dr. Giles to explain the Indo-Iranians, the Aryan inspirers of *Našili*, the Phrygians and the Armenians. Though the material at our disposal is still inadequate, the general character of the culture of the more vital region, north-western Asia Minor, is not unfamiliar. A mere inspection of the finds from Troy and Yortan suffices to show that we have there no pure extension of Danubian culture as we have seen it at Butmir and from Moravia to Belgium. In fact we have seen in previous chapters how very slender are the links which can be found to connect Anatolia and Europe in any direction. We did indeed detect at a period posterior to the purest phase of the peasant culture elements of cultural community between Troy and the Danube valley as well as Thrace and Thessaly. But none of those elements are distinctively Danubian in origin. Some of them come from Troy, while the battle-axe which is genuinely European is no more at home in the Danube valley than in Troy. Thus the requisite links between the Danube valley and the Ancient East cannot be established by way of Anatolia. We shall later see that the alternative route across South Russia and the Caucasus is even more definitely excluded.

Nor does the Danubian manner of life really correspond very satisfactorily to the primitive Aryan culture deduced by linguistic palaeontology. We shall not insist on the prominence of agriculture and swine-breeding, since Dr. Giles attributes both arts to his *Wiros*. But the absence of arrowheads or other weapons deserves notice. Again the cult of a Mother Goddess is an un-Aryan trait among the Danubian peasants. But the last word sums up the crucial objection to identifying the first neolithic inhabitants of the Central European loess lands with the Aryans. Without subscribing to the extravagances of the "racial psychologists" (p. 163), it may be said that the Danubians must have acquired a specific mentality, that of the peasant. The peculiarly sedentary agricultural culture which we have described must have stamped its authors with the essentially peasant outlook still so familiar in eastern Europe or China. That its narrow conservatism, its intense attachment to the soil should ever have developed of itself into that restless love of wandering and acquisitiveness which has not only diffused Aryan languages over half the globe, but also imposed them on so many non-Aryan peoples is highly improbable.

Of course cases are not unknown in which sedentary peoples have taken to nomadism; that might have happened to the Danubians and changed their mentality. But there is no evidence in Europe for such a climatic crisis during the geological present as could have induced the cultivators of the loess to make that adventure into the unknown. As a matter of fact the Danubians survived in Central Europe. The peasants appear, as far as we can judge, throughout the prehistoric age, as an inert mass, and have ever been the prey of a series of conquerors, just as they appear in history. Historically the peasantry have again and again passed under the rule of new lords, and often have submissively adopted the language, customs and beliefs of the conquerors. And we shall see the Danubians themselves continually being mixed with, and overlaid by alien ethnic and cultural types, even in the prehistoric period. Much that was at base Danubian was preserved and persisted as it persists to-day;

for are not the Virgins of Austria and Serbia the survivals, transformed and sublimated, of the Mother Goddess whom the first Danubian peasants modelled in clay? It is unlikely that the Aryan language, at once the product and the matrix of Aryan psychology, was the work of such peasants. We should rather expect to find the Aryans emerging after the peasants had become mixed with other more venturesome elements. It would not inevitably follow that the other elements were Aryan before they reached the Danube.

2. *The Rite of Cremation*

If we reject the identification of the Danubian I peasants with the Aryans, it still remains possible to argue that one or more of the later racial groups that settled in the Danube valley became Aryan there. This is in effect the contention of E. de Michelis.[1] He starts from a very remarkable phenomenon observed in the Danube valley and the surrounding regions during the Bronze Age—the change from inhumation to cremation. The mode of disposal of the dead is often regarded as one of the most fundamental customs of a people, and one that they would most tenaciously preserve. Yet we see the new practice of burning the body spreading into regions where the dead had previously been interred. Our author thinks that the spread of the new rite was due to a racial migration, and that the migrants were Aryans setting out from the Danube valley.

De Michelis points out, as we have done, that the rite was introduced into Italy by the *terramaricoli*, who were Aryan invaders. It was followed by the Umbrian Villanovans and the Veneti of Este coming, like the *terramaricoli*, from Central Europe. For Greece Ridgeway can be cited as bringing the rite of cremation with the Achaeans from the same quarter. In the west our author shows that the cremation graves, which during the Late Bronze and Early Iron Ages spread through France and ultimately reach Spain, may well be due to Celts coming from east of the Alps and the Rhine. In Scandinavia and North Germany the inhabitants took to burning their dead about the same time. Further east de Michelis assigns the fields of cinerary urns of the Lausitz type which extend from Bohemia to the Vistula

[1] *L'Origine degli Indo-Europei*, 1903, esp. cap. ix.

and beyond to the Slavs, and proposes to derive the whole culture from Hungary. In Hungary itself cremation is well attested at least by the Middle Bronze Age. Thus the Italian philologist can present the cremators as radiating from Hungary and show that all were Aryans. In Asia the Aryan Hindus practised cremation, and we know now that in the Indus valley that rite superseded the older practice of inhumation. Intermediate links are indeed lacking unless the change from inhumation to cremation, about 1100 B.C., at Carchemish be regarded as a reflection of the passage of Aryan cremationists. But the rite itself is a material bond.

As thus stated de Michelis' thesis achieves the finest cultural synthesis among all Aryan peoples yet found. It gives a distribution of a cultural peculiarity which harmonizes exceptionally well with the distribution of Indo-European languages. Yet on closer examination the difficulties seem almost insuperable. The facts just stated are correct, but they are not all the facts. Our fuller knowledge of 1925 reveals that cremation presents a much more complicated problem than an author writing in 1902 could imagine.

In the first place instances of cremation, earlier than those cited, have come to light. In Britain [1] burnt bones have been found in "neolithic" long barrows and again in round barrows of the second phase of our Bronze Age, which is still contemporary with the Continental Early Bronze Age. In both cases then the rite appears earlier in Britain than that expansion of Celts to which de Michelis attributed its diffusion westward. The position in Brittany is much the same. In Central Europe the phenomena are even more intricate. In the Neckar valley [2] burnt human bones have been found with sherds typical of the Danubian I peasants and some of the same people seem to have cremated in Bohemia too. Elsewhere in the Rhineland barrows belonging to a Nordic battle-axe folk occasionally cover cremated bones.[3] In North Germany [4] ashes contained in cinerary urns have sometimes been found in late megalithic graves. In Thuringia and Saxony [5] a Danubian II people using pottery and other artifacts similar to those met in the inhumation graves of Lengyel in

[1] *Dawn*, pp. 288 and 296.
[2] Wolff, "Neolithische Brandgräber der Umgebung von Hanau," *P.Z.*, i.
[3] *Dawn*, p. 257.
[4] Schumann, *Die Steinzeit gräber der Uckermark.*
[5] *Mannus*, xi–xii, pp. 312 ff.

Hungary (p. 150) had taken to incinerating their dead, while other late neolithic cultures in the Elbe valley also belong to cremationists. Finally in Moravia [1] the bell-beakers, usually accompanying the inhumed skeletons of the Prospectors from the West (p. 99), in one or two cases contained cremated remains. All these cremations in Central Europe are locally classed as neolithic or chalcolithic and are to be dated at latest between 2400 and 1800 B.C. And they were associated with material which is normally found with inhumed skeletons belonging to different racial types, Mediterraneans (Danubians), Nordics and Prospectors! To add to the confusion, isolated cases of the rite have been reported from Hither Asia at a very early date—in "neolithic" deposits at Gaza [2] in Palestine, and about 2000 B.C., in a "fire necropolis" at Surghul [3] in Babylonia (the latter very doubtful however). These scattered remains cannot be neatly linked up like the Bronze Age examples, on which Dr. de Michelis relied.

Secondly in many cases inhumation gives place to cremation without any other signs of a break in the general continuity of culture or of the presence of a new race. Thus in Britain the use of metal was introduced by short-headed invaders from the Continent, who built round barrows (as contrasted to the neolithic long barrows) but inhumed their dead. The artifacts found in rather later round barrows of Bronze II covering burnt bones give no sort of indication of a fresh invasion from Central Europe or anywhere else. Again Scandinavian archaeologists insist emphatically on the complete continuity of culture between the epoch of inhumation and the subsequent period of cremation in the Danish and Swedish Bronze Age. Indeed, the two methods of burial are often met at different levels in the same barrow. Not only so, from the exiguous skeletal remains from the cremation epoch and the richer material later available after the reversion to inhumation, it is clear that the skulls belong to exactly the same racial types as existed in Denmark and Sweden from late neolithic times when the dead were interred.[4] The cultural continuity is even more strikingly exhibited in South-west Germany. The barrows of the Nordic battle-axe folk on the Neckar and Lower Main provide a most instructive series. It had always been the custom of this people both in the Rhineland and in Thuringia,

[1] *W.P.Z.*, vi, p. 41 f.

[2] MacAlister, *Excavations at Gezer.*

[3] *O.L.Z.*, xxi.

[4] Pittard, pp. 210–12.

whence their ancestors had come, to kindle a great fire perhaps for a funeral feast in the trench destined to receive the corpse. Dr. Schliz[1] has very plausibly suggested that, after a time, the custom arose of casting the corpse on this fire without waiting for its extinction as had been originally the wont. Here is a possible explanation of the local rise of the new practice. In Bavaria cremation only comes in gradually during the Bronze Age, and to illustrate this we may cite the curious transitional observances: part of the body was interred unburnt while part was cremated, and the ashes deposited in the same grave enclosed in an urn.[2]

Thus the rite of cremation not only appears at various times, and at widely separated centres among people apparently belonging to different physical types, but also its introduction is associated with no other symptoms of racial change; the new rite develops gradually as if spontaneously and does not as a rule come in catastrophically. The extraordinary complexity of its distribution both in space and time makes the reference of the practice of cremation to a single race or an unique focus exceedingly perilous. Nevertheless the phenomenon is perplexing. There always remains the possibility that there was somewhere in prehistoric times a people who always cremated but, who because of this very fact and because they used artifacts of perishable materials such as wood and leather, are and must remain unknown to the archaeologist. Myres[3] and Christian[4] incline to the view that the brachycephalic Alpine race both in Europe and Asia regularly practised cremation. Burials by this method are admittedly very difficult to detect. It must be remembered that no adult burials are yet known in connection with several cultures—those of the vase-painters of Anau in Turkestan and of Thessaly and South-east Europe in general or of the Alpine lake-dwellers for example. The distribution of the rite in prehistoric Europe and in Hither Asia certainly needs elucidation. A thorough study of the burial rites of the Cappadocian Hittites and of the circumstances under which cremation replaces inhumation in the Punjab may reveal that this rite is intimately bound up with the solution of the Aryan problem. At the moment the identification

[1] "Die Schnurkeramische Kulturkreis" in *Z.f.E.*, 1906.

[2] Dechelette, *Manuel*, ii, p. 157. Cremations and inhumations already occur side by side in the same barrow in the Middle Bronze Age.

[3] Or rather that all cremationists were Alpine (*C.A.H.*, i, p. 73). He is certainly mistaken in attributing the early inhumation graves of Switzerland to the lake-dwellers, and the evidence for cremation among the vase-painters in South Russia is very dubious and generally rejected by the most competent authorities to-day; cf. *J.R.A.I.*, liii, p. 267.

[4] *M.A.G.W.*, liv, p. 42.

of the Aryans with the nebulous and hypothetical people who diffused it would seem at least premature and the localization of its original focus in Central Europe utterly groundless.

On the other hand, it may be doubted whether cremation is really such an ingrained and characteristic habit of a race as Professor Ridgeway and most Italian prehistorians imagine. It is alleged that the burning of the corpse implies a peculiar belief in a world of the sky whither the soul of the departed is conveyed by the funeral fire instead of descending to the underworld or abiding in the tomb. It cannot be said that a study of the funerary customs of "primitive" peoples has confirmed this doctrine. On the contrary, it has shown that a great diversity of burial rites subsists among culturally and physically homogeneous tribes.[1] In America some Indians cremate, some dispose of their dead in other ways. It is just the same among the Melanesians and the Pacific islanders.[2] In Australia, where the population is extraordinarily uniform in physical type, cremation is but one of many rites in vogue. Even within a single tribe it may be reserved to a particular class or grade while other members of the same tribe are inhumed or exposed on platforms or trees.[3] What is still more remarkable is that both inhumation and cremation were practised by the Tasmanian aborigines.[4] Yet this race had preserved a palaeolithic culture, no higher than the European Aurignacian unaffected by any foreign influence till their extermination. Such instances warn us against attaching too high a value to burial customs as criteria of race. In the case of the Siberian aborigines it is clear that burning has been adopted because no other method of disposal of the corpse was practicable, the ground being too hard for a grave to be dug.

In any case it can neither be shown that all Aryans cremated, nor that all cremationists were Aryans. In Bosnia, the earliest barrows usually ascribed to Illyrians cover unburnt dolichocephalic skeletons. The Early Iron Age graves of Macedonia, which must be ascribed either to proto-Dorians or the Dorians' immediate cousins, were invariably by inhumation. On the Greek Mainland no cremations were observed in the early Geometric cemeteries of Tiryns, Asine, and Argos. As we have remarked above it is quite as likely that the custom spread to Greece from Asia Minor as that

[1] See the article "Burial Rites" in *Hastings' Encyclopædia.*
[2] Cf. Fox, *Threshold of the Pacific*, 1924, pp. 217 and 229.
[3] *M.A.G.W.*, xlvi, p. 86. [4] Ibid., p. 84.

it was introduced from Central Europe by Achaeans or Dorians. Again the earliest Phrygian barrow at Gordion contained an unburnt body. On the contrary, neither the early cremations from Palestine nor the questionable examples from Surghul in Mesopotamia can be attributed to Aryans. And though the people of Carchemish, who began to cremate about 1100 B.C., may be called "Hittite", there is no evidence that they were admixed with the same Aryan element as had influenced the Hittites of Cappadocia three or four centuries earlier.

In the light then of this cursory survey of burial rites, ancient and modern, it may be said: (1) No single race is identifiable, either somatically or by means of its pottery or implements, to which all the cremations even in Central Europe during the IIIrd and IInd millennia B.C., let alone those of Britain or Hither Asia and modern examples in Australia and America, can be traced. (2) The change from inhumation to cremation can in some cases at least be shown to be accompanied by no change in race detectable anthropometrically or culturally. (3) It cannot be proved that the practice of burning the dead originated in and radiated from Central Europe. (4) Cremation is not universally attested among the earliest Aryan peoples, while it was sometimes practised by non-Aryans. On these grounds the attractive hypothesis of Dr. de Michelis as stated above must be abandoned.

3. *The Nordics in the Danube Valley*

The possibilities of Central Europe are still far from exhausted. In the Danube Valley other cultural groups grew up and expanded upon the foundation prepared by the Danubian peasants. In Hungary and Moravia the Danubian I culture gives place about the middle of the IIIrd millennium B.C. to a new group,[1] centred in Hungary conserving many of the old elements with an infusion of new ones. Socially the unit of organization is enlarged and at the same time consolidated, and opposed to other groups; villages are now more extensive, large cemeteries are laid out near them, fortification walls are sometimes thrown up, weapons of war are manufactured. Culturally, the innovations have a double aspect; on the one hand a sort of barbarous vase-painting, in which the coloured designs are laid on the surface of the vase after its polishing

[1] Described with illustrations in *Dawn*, pp. 176–80.

and firing, and certain ceramic types, such as pedestalled bowls, point to an impulse from the Transylvanian vase-painters to the east; on the other, certain vases and spectacle spirals of copper betoken contact with Troy II. Ethnically a change is denoted by the presence of tall dolichocephalic skeletons of Nordic type in the cemeteries of Bodrogkeresztur near Tokay and of Lengyel in Tolna County, south of Budapest.

Now some authors, who hold that the Aryans belonged to the Nordic race, yet consider that they formed only one branch of that race. De Lapouge[1] placed the area of characterization of the Aryan branch in Central Europe, whither the Nordics

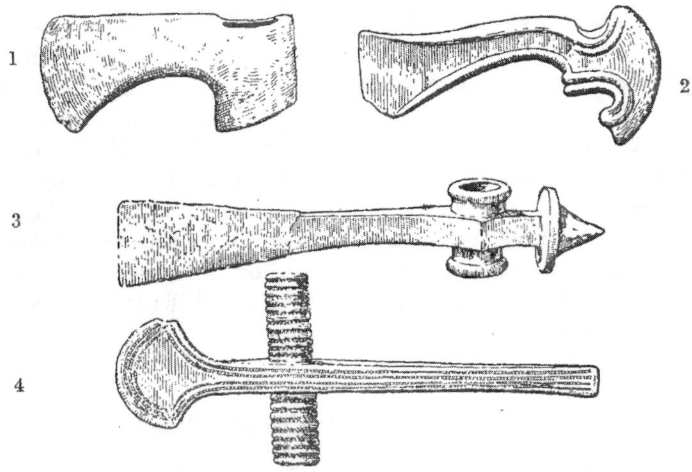

Fig. 24. Hungarian Battle-axes. 1, Copper Age; 2–4, Bronze Age.

would have penetrated by a gradual infiltration. Now here in Hungary we have evidence of this infiltration. And the culture possessed by the Nordics of Hungary, shows affinities with the contemporary cultures of Illyria, Thessaly and Troy. Its ramifications can further be traced into Moravia, Silesia, Bohemia, Saxony and Bavaria. We thus have in the second Danubian period external relations which were lacking in period I. Are not the conditions postulated by Dr. Giles now fulfilled? I hardly think so.

The Danube valley at this epoch does not seem so much an original focus from which culture radiated as a secondary centre

[1] *L'Aryen*, 1899; cf. *R.E.A.P.*, 1887.

where new elements, derived from without, were elaborated. The connections with Troy, for instance, do not illustrate an influence upon Troy so much as one from Troy. At the same time the most European elements in Troy are not lineal descendants of any Danubian elements. The elaborately shaped stone battle-axes of the Troad do not seem to occur in Hungary at this date, but the first copper battle-axes do begin to appear. We get the impression that we are looking at two sides of the same phenomenon. There was perhaps an infiltration of Nordics through Bulgaria into the Troad, marked principally by stone battle-axes, and a distinct infiltration into Hungary, marked by the dolichocephalic skulls and the copper battle-axes. If the latter movement reached Hungary from the East across the Carpathians, it might have brought in its wake those elements of Transylvanian culture which appear barbarized in Danubian II. But in that case, and if the Nordics be Aryans, there is no reason for restricting that denomination to the invaders of Hungary alone since the latter did not advance to Asia Minor. They were rather but one branch of a single Nordic migration, the centre of which lay outside the Danube valley. And it would be to that centre that we should look for the Aryan cradle.

The justice of this view is demonstrated in the succeeding period, when an invasion of Nordic peoples can be traced beyond all doubt in Moravia and Bohemia. The intruders who pour in round about 2000 B.C., in several bands bring with them, ready made, a complete apparatus of civilization the prior evolution of which can be traced in detail further north or east. Rugged hunters and herdsmen conquered the lands occupied by the Danubian peasants and, unlike these, established themselves by preference on hill-tops. These mobile tribes brought with them new types of vases[1]—the so-called Nordic pottery with the first wave, cord-ornamented ware with the second—strange implements, flint instead of stone celts and splendid battle-axes of stone, and sometimes heaped a barrow over their dead. Similar incursions were reaching Transylvania and Hungary about the same time and left as their monuments barrows, fresh copper battle-axes and stone ones as well, and sherds of corded ware shown by their distribution to have come from the south-east and east as well as from the north.[2] The Danube valley was thus occupied by

[1] *Dawn*, figs. 112–16.
[2] *Dologozatok*, vi, 1915, pp. 1 ff.

Nordics, and not the centre from which any group of Nordics moved northwards.

Now these Nordic invaders must already have been Aryans if the Aryans were in truth primarily Nordic; for the subsequent cultures both in Hungary and Bohemia, in so far as they were Nordic, developed out of the intrusive cultures of the third period, just described. After this date there was no further intrusion of peoples till well on in the Late Bronze Age. The culture of the Danubian Bronze Age is essentially continuous with that of the last neolithic or chalcolithic epoch. The pottery of the Early Bronze Age or Aunjetitz [1] graves of Bohemia, Saxony, Silesia and Moravia is derived from the Nordic pottery crossed with types associated with the Prospectors and with the earlier Danubian II culture of Lengyel; battle-axes of stone, horn, or rarely bronze are found in Aunjetitz graves, and in eastern Hungary the typical weapon of the Bronze Age is the metal battle-axe. The introduction of regular metallurgy had not been accompanied by any further addition to the population. Traders had indeed arrived—the Prospectors who brought the bell-beakers from the west were one small band—but the stimulus in industry was supplied by trade with Troy in tin and with Crete and Greece in amber. But the people who used the metals and worked them belonged on the whole to the earlier stocks. By the latter half of the second millennium they had outgrown the leading-strings of Asia or the Aegean, and created an original series of forms; above all they evolved a superior weapon, the slashing sword, which was destined to subdue the Aegean and then the Ancient East, which hitherto had known only the rapier or the dirk (Fig. 25).

Hand in hand with these cultural transformations had gone social revolutions which led to the emergence of aristocracies of war superimposed upon the old peasant communities. In contrast to the peasant art of the Stone Age, the Danubian Bronze Age art bears the stamp of a barbaric chivalry. At the same time the social structure had been enlarged to embrace a wider horizon than the village and concomitantly therewith chieftainship and sovereignty arose. The predominance of individuals is plainly attested by the Early Bronze Age, when royal barrows, furnished with a wealth of gold ornaments and princely weapons, were raised beside the simple flat graves of the plebs. Very likely this concentration of power dates back

[1] *Dawn*, pp. 191–200; *Arch.*, lxxiv, p. 164, fig. 8.

to an earlier period when the Nordic invasions began. The battle-axes which then appeared, like the ceremonial axes of Hissarlik, may well have been emblems of authority, and there are some indications that the rugged pastoralists who at that time occupied the hills established themselves as overlords among the older and simpler peasants. Through the clash between sedentary

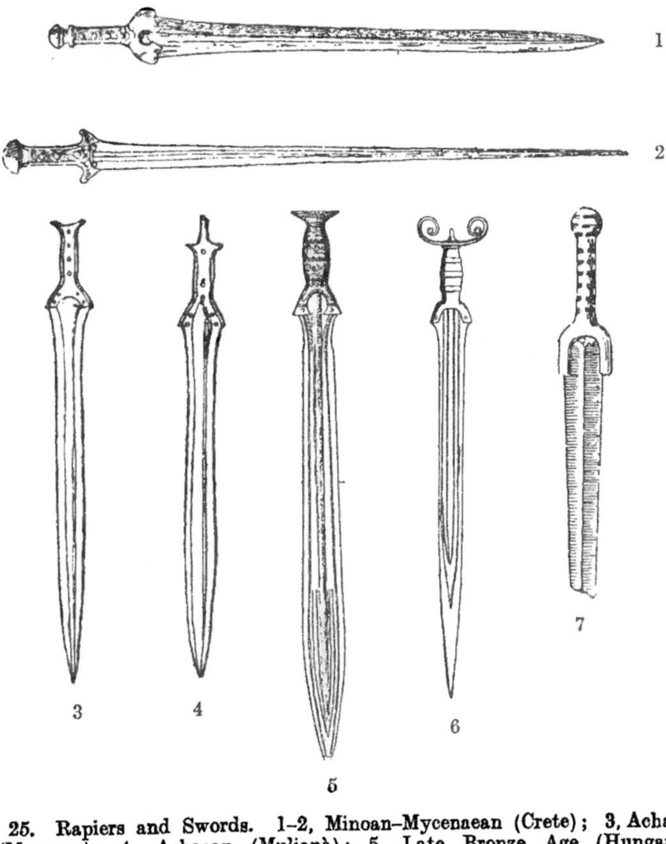

FIG. 25. Rapiers and Swords. 1–2, Minoan–Mycenaean (Crete); 3, Achaean (Mycenae); 4, Achaean (Mulianà); 5, Late Bronze Age (Hungary); 6, Antennae-sword (Switzerland); 7, Mouçi Yeri (Transcaucasia).

and nomadic peoples and the social convulsions which ensued the rigidity of the agricultural communities had been broken down, the basis of life widened and the way paved for the sudden burst of industrial and artistic activity that culminated during the Bronze Age. In the ferment which produced this result one would feign see the work of the Aryans. But before these can be identified with the tall Nordics whose advent we have described, another contemporary band of invaders must be considered.

4. *The Alpines Descent upon the Danube Valley*

During the period of disturbance other peoples had descended upon the peasantry of the loess lands of Central Europe. They were apparently mountaineers coming from the highlands of the west and may be called Alpines without prejudice to any theories about their skulls. It must be asked whether they were not Aryans.

The original habitat of the new-comers must have been the highland zone, and they were themselves the descendants of that pre-neolithic people whom we came to know in the last chapter. They had inherited, as we then saw, the habit of building their habitations on rafts or piles, on the shores of lakes (p. 126), and had acquired some of the neolithic arts from the Danubian peasants, whose territories they now were to invade. Thus equipped they created the well-known neolithic civilization of the Swiss lake-dwellings, while others among them in the Rhine valley, Baden and Bavaria, built fortified settlements on hill-tops and evolved the cultures called, after the type stations, Michelsburg and Altheim respectively.[1] These neolithic cultures were belated in comparison with Danubian I, but their authors, at once pastoralists, agriculturalists and hunters, had made great advances in social organization. The erection of pile or platform villages upon the lake shore would involve co-operation in social labour. The land settlements were fortified with a ditch and a moat and the huts arranged in regular rows, all of which implies a collective direction and a deliberately co-ordinated effort. This definitely social character may, like the architecture which embodies it, have been a heritage imposed upon the Alpines by the stringent conditions of life in pre-neolithic times.

By the third period of culture in the Danube valley, the creators of this highland civilization, although split up into a multiplicity of cultural groups, began to expand. That was inevitable. The population of the narrow mountain valleys has always been overflowing, whether as raiders, mercenaries or waiters. So stray

[1] *Dawn*, pp. 252–8.

groups of the prehistoric hillsmen from South-west Germany reached Bohemia [1] about the same time as the Nordics. From Switzerland a descent was made upon the lakes of Upper Italy. Ceramic remains from several sites in Lower Austria and probably Moravia too betoken the presence of a branch of the same stock come from the East Alpine slopes. In the Middle Danube area pile-dwellings were planted, at an as yet undetermined date, along the banks of the Danube and the Theiss and on the lakes of Carinthia, Carniola, and Styria. By the Late Bronze Age the habit had spread to Bosnia and the classical authors related how in their days such structures were inhabited in Macedonia.

Now there are good reasons for considering this Alpine zone as in a sense pre-eminently Aryan. Thence started the Celts from somewhere between Thuringia and Istria. The Italici emerge on its southern fringe and the structure of their *terremare* is evidence that some of their ancestors had lived in pile-dwellings (p. 71). Again there is an unmistakable affinity between the pottery of the Bosnian lake-dwellings and that of the *terremare* on the one hand and of the Early Iron Age of Macedonia on the other. The latter can be regarded as proto-Dorian, if not proto-Hellenic, just as the former has been classed as Italic. Incidentally the same types, among which curiously elaborated handles are typical, can be traced far across South Russia to the Dniepr and beyond in the full Iron Age in Scythian barrows.[2] Remoter parallels are quotable from all parts of Hungary by the Middle Bronze Age, and persist into the Hallstatt period. Thus a considerable number of the Aryan races in Europe can be connected with the group of Alpine cultures. Even wider connections could be established if we could accept Taylor's view [3] that the physical type of the original Aryans was that of Ridgeway's Celts, tall, blonde and brachycephalic, one branch of which stock is represented by the British round-barrow men; for the type may have been evolved in the highland zone of Europe.

It is not, however, really easy to establish wider cultural connections from the highland zone to the east or the north. In the first place no positive assertion can be made about the prevailing burial rite; for no early burials belonging to the lake-dwellers of Switzerland, Upper Austria, Carniola or Bosnia are known,

[1] Ibid., p. 185.
[2] *Izvestia. Imp. Arch. Komm.*, xxxv, pp. 66 ff., figs. 2, 16, 19.
[3] *The Origin of the Aryans*, pp. 105 f.; see below, p. 161.

and the same remark applies to the inhabitants of the Bavarian land-stations of the Altheim type. On the other hand the allied Michelsburg people in the Rhinelands and even in Bohemia always buried the dead unburnt in pits under their huts, which were then destroyed. It is not therefore permissible to claim these Alpines of Central Europe as the original cremationists whose existence was envisaged as a possibility a few pages back. At the same time all these Alpine cultures belong to a relatively late date, and have borrowed largely from other groups, so that it is hard to say how much of their content is original.

Nor does Alpine civilization in its earliest phase correspond well with that deduced for the Aryans. Highland country broken up by valleys and precipices is not where one would look for an early acquaintance with the swift horse, a pre-eminently Aryan animal. And as a matter of fact his bones have not been found in the oldest stations in Switzerland. On the contrary we should be inclined to look for some evidence of the Aryanization of the Alpine zone.

Now among all the peoples of the European highland region, there is evidence of a Nordic admixture and that anterior to the period of their demonstrable expansion. Beside the Michelsburg settlements in Wurtemburg and Baden often rise barrows, containing cord-ornamented pottery and battle-axes, that may mark the sepultures of Nordic chiefs. In the Bavarian Altheim settlements and in the villages on the Upper Austrian lakes, numerous stone battle-axes have come to light (cf. Fig. 27, 5) the history of which is to be sought further north and east. Here, as at Troy, they may be the symbols of the authority borne by Nordic chiefs. Similar axes occur moreover in Upper Italy[1] both in the *terremare* and before. The pottery from the Michelsburg hill stations and land stations in Lower Austria again includes types reminiscent of the Nordic fabrics which we have met intruding into the Danube valley, and the oldest lake-dwelling on Laibach Moor in Carniola actually yielded a cord-ornamented beaker. The crescent-shaped handle itself, so distinctive of the Italian *terremare* and the late Bosnian lake-dwellings, may be derived from or at least influenced by earlier or simpler types appearing in a Nordic context in Bohemia, Moravia and Galicia. The porched house of the Michelsburg folk which so strongly recalls a Greek megaron (Pl. VIII, 1) is considered by Reinerth to be of Nordic

[1] *Dawn*, p. 266.

origin.[1] Thus on the eastern slopes there is abundant cultural evidence for contact between the Alpines and the Nordics. In Switzerland itself craniological evidence is also forthcoming; not only do we find intrusive barrows with pottery and artifacts of Thuringian and Rhenish types, penetrating as far south as Canton Zurich, not only do the later pile-settlements even on the Lakes of Geneva and Neuchâtel contain Nordic battle-axes of stone, but further, characteristic dolichocephalic skulls from such later villages bear witness to an infiltration of Nordics among the originally brachycephalic Alpine population. As we shall shortly see, many anthropologists are inclined to regard the tall brachycephals as the results of crossing between the Alpine and the Nordic stocks.

Hence in the late neolithic period a Nordic element among the inhabitants of the highland zone must be admitted both on cultural and anthropometric grounds. We have already seen that such an element constitutes the bond of union between Central Europe and other areas. Therefore it will perhaps not be overbold to see in it, also in the Alpine regions, the source of the Aryan element. That is, we might regard the Alpines as Aryanized by admixture with or conquest by these Nordics.

A whole chapter in the pageant of European prehistory has now been unfolded before us. We have witnessed the opening up of the Central European plain to food production by simple peasants, the Danubians. We have watched rude hunters in the highland zone, acquiring neolithic arts from these Danubians and founding the cultures of the lake-dwellings and hill stations. We have traced upheavals and invasions with the aid of rude artifacts left by nameless peoples and seen how such invasions affected the highlands and the plain alike. And we have observed a most significant change—the transformation of a peasant culture, diffuse but essentially conservative and unprogressive, into the nucleus of a civilization pregnant with potentialities of development and expansion. Out of the clash of diverse cultures and different racial elements with contrasting economic organizations and social traditions, the barbaric rudiments of States were generated. Within the skeleton of these, we discern the womb, fertilized by trade with south and west, whence at least two historic nations, the Italici and the Celts, came forth to conquer

[1] H. Reinerth, *Der Wohnbau der Pfahlbaukultur*, 1924, pp. 11 f.; for a criticism of this view see Boethius in *B.S.A.*, xxiv, pp. 161 ff.

and Aryanize a large part of the continent. And as the driving force therein we could recognize one element which we also met at Troy and in the Aegean. Thus, though we have not found the Aryan cradle, we have a clue which guides us to that region upon which linguistic and ethnological data seem alike to converge, the great plain of North and East Europe. On it the Nordic race was admittedly characterized and thither we now repair.

CHAPTER VII

THE THEORY OF A NORTH EUROPEAN CRADLE

1. *The Aryans as Blondes*

The great majority of investigators from Omalius de Halloy and Latham onwards, who have accepted the doctrine of an European cradle-land, have located it somewhere on the great plain that extends from the North Sea to the Caspian. Not only does this region fulfill the conditions postulated by linguistic palaeontology better than any other, it was also the area of characterization where the tall blonde stock, the European race *par excellence*, was evolved.[1] And all advocates of a cradle in Europe who have appealed to anthropological results at all, have conceived of the original Aryans as blondes.

The pioneers of the European theory devoted much ingenuity to showing that peoples of tall stature, fair hair, light skin, ruddy complexion are or were to be found wherever Aryan languages are or have been spoken. They had, of course, to concede that to-day these physical characteristics have almost entirely vanished among many peoples who are linguistically Indo-European. Even in Europe the typical Greeks, Italians and Spaniards are short, dark Mediterraneans. But such phenomena can be explained by racial admixture. The tall blonde Aryans in such areas would have been only a conquering minority. Their physical characters, evolved in a cold climate, had only a low survival value, were recessive to use the Mendelian terminology, in the new environment. But much evidence could be adduced to show that in regions linguistically Indo-European, where blondes are now virtually extinct, such types had existed in antiquity.

De Lapouge [2] for example pointed out how the Achaean heroes are described as ξανθοί,[3] that among the Romans, Sulla, Cato and others seem to have been fair, while such names as Ahenobarbus, Fulvus, Flavius and Rufus imply features in complete contrast to the typical Mediterranean. In Indian literature the word

[1] Haddon, p. 151. [2] *L'Aryen*, pp. 187 ff.

[3] Cf. Giles, *C.A.H.*, ii, p. 22. But even though this word may not mean "golden" haired, it does imply a contrast to the dark Mediterraneans.

for caste is *varṇa*, "colour," already in Vedic times, and since the Pandavas in the later epics are described as tall and fair, it may be concluded that the distinction intended was that between swarthy Dravidians or pre-Dravidians, *dasyus*, described even in the Rigveda as black-skinned (*krişṇavarṇa*), and fair Aryans. We have already noticed how white-skinned slaves are mentioned in Babylonian documents just about the time when the appearance of the horse and the advance of the Kassites suggested the presence of Aryans on the Iranian tablelands (p. 24). From a study of the Persians depicted on the sarcophagus of Sidon and other monuments, de Ujfalvy [1] deduced that this branch of the Iranian race included Nordic blondes. Chinese writers describe ruddy complexioned, blue-eyed, fair-haired peoples in Central Asia at the time of the maximum dispersion of Iranian speech just before the beginning of our era, and ancient paintings from the buried cities of the Tarim basin [2] depict distinctively European blonde types beside the native Mongoloids. The Iranian Alans again were tall and fair according to the Roman writers. But in addition to these vestiges of genuine blondes of seemingly European type in the Aryan zones of the Ancient East, the same type still survives to-day among various peoples of Indo-European speech in parts of Asia where they have been sheltered by remoteness from racial admixture or favoured by a climate comparable to that of Europe. Such blondes are the Iranian Galchas first studied by de Ujfalvy in 1878. The Kurds of the highlands north of Irak are again tall, fair and blue-eyed (Pl. VIII, 2), and von Luschen [3] has drawn attention to the coincidence between their habitat and that of the Indo-Iranian Mitanni chiefs 3,500 years ago. Again round the ancient Persian capital of Persepolis a few individuals with blue eyes and chestnut-coloured hair [4] seem to betray at least an infusion of genuine blondes in Iranian Persia.

These examples ancient and modern could be multiplied, but enough has been said to show that the blondes constitute a racial link of the kind which has been sought between Europe and Asia and Northern Europe and the Mediterranean. If we accept this racial link as identical with the linguistic, the theory of an Aryan cradle in Europe receives confirmation, since everyone

[1] *L'Anthr.*, 1900, pp. 23–56 and 193–234.
[2] Feist, p. 498 and frontispiece.
[3] *J.R.A.I.*, xli, pp. 242–3.
[4] Haddon, p. 102.

NEOLITHIC MEGARON-HOUSE FROM WÜRTEMBURG, RECONSTRUCTED BY THE URGESCHICHTLICHES FORSCHUNGSINSTITUT, TUBINGEN

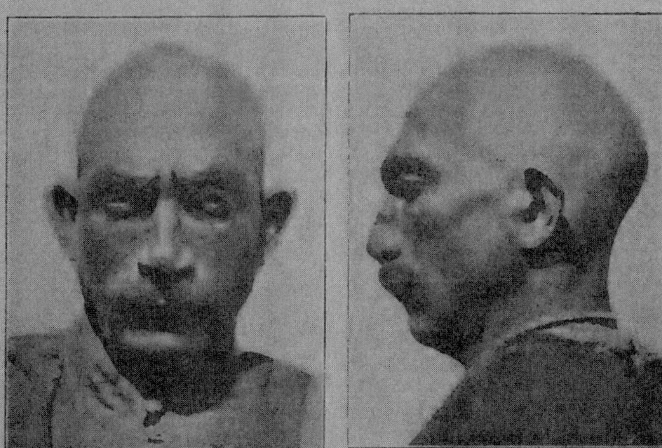

KURD

[*face p. 160*

agrees that such blondes were characterized somewhere in the European area.

Most partisans of that hypothesis who have gone thus far go further and regard the original Aryan type as that of the blonde race par excellence, *Homo europaeus nordicus*, the tall dolichocephalic Nordic. To this, however, there are notable exceptions. Canon Isaac Taylor[1] pointed out two facts; not all Nordics can be regarded as Aryan—many Finns belong to that stock—and many Aryan blondes are and were brachycephalic. To the latter type Taylor assigned the British round-barrow men who were certainly brachycephalic and very likely both Aryans (Celts) and fair. Starting out from this British material Taylor contends that the first Aryans were tall, blonde, but short-headed. This is the type represented among the ancient Celts and the modern Slavs[2] and he detects it among the Umbro-Latini too. As is well known Professor Ridgeway has elaborated the argument in the case of the Celts and extended it to embrace the Achaeans of Greece as well. Recent research has shown that the Cretan Sphakiotes, who may be pure descendants of the Dorians, belong to the same brachycephalic group.[3] Furthermore some of the Asiatic blondes[4] who have been welcomed with such éclat by the anthropological supporters of the Nordic hypothesis turn out to be as distinctly brachycephalic as the Slavs. Finally even in Germany, Russia and Scandinavia, the very regions whence the believers in Nordic Aryans derive their most powerful arguments, a brachycephalic element existed in neolithic times.

Thus with the evidence at his disposal, Taylor made out a very strong case for brachycephalic Aryans, and the later results which I have here inserted only strengthen his position. His conclusions were briefly as follows. The European brachycephals fell into two divisions, short and dark on the one hand, tall and fair on the other. Both branches were in the last resort Asiatic in origin. The former division would include the ancestors of Finns and Basques, the second would have become specialized in northern central Europe and with its complexion would have changed its speech, becoming Aryan. On the other hand the Nordic long-

[1] pp. 231 ff.

[2] Taylor describes the Lithuanians too as brachycephals, but this seems a mistake. Cf. p. 167, note 3.

[3] *B.S.A.*, xvi, pp. 257 ff.

[4] E.g. the Galchas, but not the Kurds.

heads were not originally Aryans, but only became Aryanized through contact with the brachycephals. Taylor points out with especial satisfaction how on the lips of Nordic Teutons the Aryan tongue was distorted,[1] witness the celebrated soundshifts (p. 7). It is only the last point which concerns us here, the details of Taylor's theory, the Asiatic origin of the brachycephals, their kinship with Mongols, the relation of Finnish and Basque, etc., are all somewhat antiquated, as the reader will perceive.

Nevertheless even the kernel of Taylor's thesis is, if not unsound, at least uneconomical. On the one hand we wonder why the Asiatic invaders should only have begun to speak Indo-European in Europe. On the other hand most anthropologists now incline to regard the tall blonde brachycephals, Denniker's Vistulans and the round-barrow men who invaded Britain, as the product of hybridization between Nordics and darker brachycephals.[2] In fact Taylor himself has to admit actual contact between tall dolichocephals and tall brachycephals in the period of the Aryans' co-existence. Is it not more reasonable, asks Zaborowski,[3] to ascribe the change, admitted by Taylor, from pre-Aryan to Aryan speech precisely to those superadded dolichocephals? An affirmative answer would not be necessary, could a specific culture be ascribed to the tall blonde branch of the brachycephalic stock. But these appear as a relatively homogeneous mass only in a late stage of European pre-history and no peculiar culture can be assigned to them. Their artifacts and burial rites everywhere show mixed characters and the element common to all is, in each case, that normally associated with Nordics. That is most conspicuously true of the invaders of Britain.[4] Their pottery and some of their metal utensils show the influence of the Prospectors. On the other hand their typical vase was not the bell-beaker (p. 99) but rather a cord-ornamented beaker decorated in the style of a bell-beaker; they used stone battle-axes and buried under barrows and not in flat graves like the Prospectors. Now all these features, the cord-ornament, the battle-axes and the barrow to mark the grave, belong to a purely Nordic stock, well-known in Thuringia and South Germany (p. 174). It is to this folk that our round-barrow men must owe their tallness and the Nordic peculiarities observed in the con-

[1] pp. 231–2.
[2] Haddon, pp. 27 and 29.
[3] *Rev. Éc. Anthr.*, 1898.
[4] *Dawn*, p. 293.

formation of their skulls as well as many elements of their civilization.

We may then regard the tall brachycephalic blondes as ethnically [1] and culturally mixed. We may accordingly identify that ingredient to which they owed their tallness and fairness and such a large element of their culture too with the Aryanizers who taught the brachycephals Indo-European. To the Nordics we now return, but perhaps in the light of the foregoing discussion we should add the proviso that, at least by the time of the Aryan dispersion, it is improbable that the Nordic or any other stock was strictly pure.

Now certain arguments extraneous at once to anthropology, archaeology and philology have been adduced to fortify their claim. The pioneers of the Nordic hypothesis and many of their disciples have ascribed to the Nordic race as such a physical superiority corresponding to the linguistic pre-eminence of Indo-European speech and have sought to deduce from the skeletal build of the Nordic the psychological characters which they regard as peculiarly Aryan. Penka in Germany and de Lapouge in France waxed lyrical in praise of the virtues of the tall blondes, and these panegyrics are still echoed by more popular writers in this country, Dean Inge for example, and above all in Germany. According to Penka the Nordic race was "ever-conquering and never conquered", it was "spiritually and physically aristocratic". A passage of de Lapouge's eloquence is worth quoting:

> "La superiorité sociale de l'*Homo Europaeus* s'accuse de toutes façons. Il occupe les plaines laissant les hauteurs à l'*Alpinus.* Il afflue dans les villes, dans les centres d'activité, partout où il faut plus de decision, d'énergie. Plus une couche sociale est élevée, plus on le rencontre en grand nombre. Il prédomine dans les arts, l'industrie, le commerce, les sciences, et les lettres. Il est le grand promoteur du progrès." [2]

It seems to-day unnecessary to quote further from the rhapsodies of those who have been called the "anthroposociologists" or to criticize their premises. The correlation between cranial contours and intellectual characters, if any, has yet to be discovered. No serious anatomist to-day would attempt to deduce from a

[1] Recent researches suggest that though the factors determining the inheritance of skull-form are exceedingly complex, brachycephalism tends to be a dominant character in a cross between long heads and short heads.

[2] *L'Aryen*, p. 399.

skeleton the spiritual aptitudes or achievements of its one-time owner. The measure of truth which underlies such fables must await exposition in a later chapter. As a contribution to the identification of the Aryans the fantasies of the anthroposociologists are quite worthless.

Not only are they worthless; they are positively mischievous. They have induced their votaries to postulate all sorts of migrations, for which there is as yet not a particle of evidence. To buttress the Nordic's claim to be the ruling race *par excellence*, attempts have been made, and are still being made, to prove that the earliest dynasties of China, Sumer, and Egypt were established by invaders from Europe and even to-day the vision of certain prehistorians is absolutely distorted by this preconception. Such misdirected enthusiasm also injures science in another way. The apotheosis of the Nordics has been linked with policies of imperialism and world domination: the word "Aryan" has become the watchword of dangerous factions and especially of the more brutal and blatant forms of anti-Semitism. Indeed the neglect and discredit into which the study of Indo-European philology has fallen in England are very largely attributable to a legitimate reaction against the extravagancies of Houston Stewart Chamberlain, and his ilk,[1] and the gravest objection to the word Aryan is its association with pogroms.

2. *Scandinavia and the Germanist Hypothesis*

Having then agreed that the original Aryans belonged essentially to the Nordic race and that the latter was characterized on the North and East European plain, it remains to localize the cradle land. Cuno (1880), Zaborowski (1898) and others have indeed argued that the whole region from the North Sea to the Caspian should be looked upon as the continuum in which Aryan language developed. That, however, seems impossible. The primitive language appears to have been too nearly an unity to have been formed in such a vast and diversified area (p. 12). Again the Aryan people were sufficiently closely knit to have a tribal god and father of their own; it is scarcely conceivable that a "tribe" or a series of tribes or families, diffused indifferently

[1] Lothrop Stoddard, *Racial Realities in Europe* (1924), imports this false principle into American politics.

over thousands of miles of marsh, forest and steppe, should have possessed the degree of coherence which this and other traits in the primitive Aryan culture imply. It must not be forgotten that for prehistoric man forest tracts, denser then than to-day, offered serious obstacles to intercourse and locomotion. While it is possible that in pre-neolithic times stray families of proto-Nordic hunters wandered over a large part of that immense plain, in the search for the Aryans it is clear that we must look for a more restricted area where a homogeneous culture was evolved and whence its diffusion can be traced.

Now two points on the plain have from the very first attracted the attention of philologists; of the two pioneers of the European hypothesis, Omalius d'Halloy selected North-central Europe, and Latham, Volhynia and the Ukraine. They showed extraordinary prevision; it may at once be said that in the present state of our knowledge the cultural conditions are fulfilled only in one of those two directions. Yet many other points have been singled out both by philologists and anthropologists. Poesche in 1878 hit upon the sources of the Pripet, the Rokitno Swamp, as the most likely spot. He was guided by a mistaken anthropological conception; confusing the Nordics' blondeness with albinism and erroneously believing that the latter abnormality is due to a marshy environment, adding that it was at his time common around the Rokitno Swamp, he located the original home of the Nordic race and so of the Aryans there. All his premises being wrong, his theory is of only academic interest. In any case no cultural group is known that originated around the head waters of the Pripet. The same latter defect is inherent in theories which set the cradle in East Germany (Hirt) or Poland. No neolithic or Early Bronze Age culture originated in either of those regions; we can clearly trace other cultures coming thither from the Danube valley, from Scandinavia and possibly from South Russia, but none of these crystallized out into an independent local culture of sufficient antiquity and importance to fulfil the requirements laid down for the Aryan cradle.

The region between the Nieman and the Vistula recently proposed by Professor Bender[1] seems to have played an equally secondary rôle in prehistoric times. It has indeed the special advantage of being inhabited to-day by the Lithuanians, a people

[1] Op. cit., p. 55.

who have preserved Aryan speech across the ages with quite exceptional purity and who belong to the Nordic race. At the moment this territory is but little explored, and some recent researches [1] have suggested the possibility that it was more like a centre of culture than has been generally supposed. Nevertheless the balance of evidence available to-day suggests that, densely wooded save for a few dunes, the haunts of pre-neolithic hunters, these regions received neolithic civilization but late, and were in no sense centres of population. Culture and presumably colonists came thither with the stamp of long development already upon them either from Scandinavia or South Russia. It is in fact to one of those areas that, in the present state of the evidence, we must turn to seek our Aryans. The former certainly and the latter probably were centres of population before the dawn of the New Stone Age, and early developed autochthonous cultures, and from them civilization and civilizers were radiated far and wide.

At the present moment the Scandinavian theory is the most attractive, having been expounded with a wealth of detail and a complete mastery of the archaeological data by such profound students as Kossinna, Schliz and Schuchhardt. The founder of the Germanist school, as we may term the advocates of a Scandinavian cradle for the Aryans, was Carl Penka. He appealed at once to history, philology and anthropology in support of his then heretical views. Anthropologically the Nordic race was traceable in the earliest remains of human habitation in the North; it is represented there at all epochs of history and prehistory, and to-day the Scandinavians preserve the type in a purity nowhere else to be equalled. In other regions such as South Russia, where the skulls from ancient *kurgans* were predominantly Nordic, or as in the Mediterranean lands and India where language demonstrates the presence of Nordics, the primitive type has given place to brachycephals or Mediterraneans. Hence only in the North, where alone the conditions for its survival have been found, could its area of characterization have lain.

The history of Scandinavia again is said to be continuous. From the time when it was first occupied by pre-neolithic men after the retreat of the glaciers, there is no trace of any foreign conquest

[1] Leon Kozłowski, *Młodsa epoka kamienna w Polsce* (1924). Rough flint implements of the type known as "Campignian", belonging generally, but not exclusively, to a pre-neolithic period, are common in Lithuania and evolve locally to neolithic forms.

or invasion. At the same time the North and not Asia has been the veritable *officina gentium.* From the beginning of recorded history we see young peoples—Teutons and Cimbri, Goths, Langobardi, Burgundians, Normans—pouring down from the cold lands to conquer and rejuvenate the effete Roman Empire. The formation of the Celts, Romans and Greeks should be regarded in the light of that analogy, the whole of prehistory will become the record of the successive swarmings south, east and west of Aryans cradled amidst northern snows. To strengthen his argument Penka did not hesitate to appeal to Homeric myth, citing the Cimmerians shrouded in continuous night and the tall Laistrygones on whom perpetual day shone as Greek reminiscences of their subboreal home.[1]

Linguistically Penka gallantly maintained that the Indo-European phonetic system was preserved in a purer form in Teutonic than in any other Aryan tongue.

The general effect of these arguments, despite exaggerations in secondary points, is undeniably very powerful. The greatest weakness lay on the linguistic side. The thesis that Teutonic is the purest Indo-European language is quite untenable; it is enough to point with Fick, Taylor and Bender [2] to the celebrated sound-shifts. Such phonetic dislocations imply that the Teutons were much mixed with non-Aryan blood. But lest that be used, as it is by de Michelis, to dissociate the Aryans from Nordics altogether let us recall that one of the purest of all Indo-European tongues is that still spoken by the Nordic [3] Lithuanians. But not only is Teutonic manifestly degenerate from a phonetic standpoint, Scandinavia and the culture of its earliest inhabitants do not correspond very satisfactorily to the picture drawn with the aid of linguistic palaeontology of the earliest homeland and primitive civilization of the Aryans. Scandinavia is essentially a maritime region and the earliest men there dwelt on the shore and lived by fishing. While the negative argument that the sea is not named in the Indo-European vocabulary is of doubtful validity, it is notorious that early Aryans even in a maritime region eschewed a fish diet (p. 84). Amber was early and universally used all along the Baltic coasts and in North Germany at a very remote date, yet no name for the gum exists in the

[1] Let us remember that Ridgeway too draws similar though less far-reaching conclusions from these passages in the Odyssey (*Early Age*, p. 358).

[2] Op. cit., p. 49.

[3] Haddon, pp. 64–5.

Indo-European languages. Though the early presence of the horse in the North has now been demonstrated, he was the small stout forest horse to which the epithet swift was scarcely applicable (pp. 83, 88). Finally, while there seem to be Aryan words at least for copper, the knowledge of metal only reached Scandinavia late and the Germanists all hold that the expansion of the Aryans began while a purely stone-age culture still reigned among them. Accordingly Professor Kossinna is obliged to regard *ayos* as a loan word borrowed after the dispersion (cf. p. 79 above).

As a matter of fact the disciples and successors of Penka have tended to dispense with the support of linguistic palaeontology; they rely on different arguments which seem to them so convincing that extraneous help is unneeded. They contend that all the Aryan races of history can be traced back to a centre on the Baltic with the aid of archaeology and that this is the only possible common focus of Indo-European speech. For the examination of their contentions it is convenient to begin with the system of Professor Gustav Kossinna [1] which may be regarded as in some respects the most authoritative. I shall then endeavour to set forth succinctly the theory which he has stated with such a profound mastery of the archaeological material, but I shall omit the mass of very highly technical detail which makes his writings so perplexing to the layman.

When the retreat of the last glaciers rendered northern Europe habitable, some descendants of the palaeolithic reindeer hunters from the West settled, about 10,000 B.C. on Kossinna's chronology, upon the shores of a series of lakes filling the depression which subsequently became the Baltic. Their chief centres would have been in Scandinavia, but relics of similar tribes are found from Holderness in Yorkshire to Latvia. These people, called by Kossinna the *Dobbertin* [2] folk but better known in this country by the name of *Maglemosians*,[3] were still just hunters and fishers like their ancestors of the Old Stone Age. Like the latter they used bone and horn very largely and worked these materials very skilfully, but they also employed tiny flint flakes, what are commonly termed microliths, for arming harpoons and many other purposes,

[1] It has been stated at length on three occasions—in *Archiv für Anthropologie*, 1902, in *Mannus*, 1910–11, and in an unfinished monograph entitled *Die Indogermanen* in 1921. In each restatement sweeping changes have been introduced so that it is hazardous to infer from the *Mannus* article how *Die Indogermanen* is going to be completed. Here the latter is followed as far as it goes.

[2] After a site near Kiel.

[3] After the site, Maglemose near Mullerup, in Denmark.

and possessed further hatchets and picks of chipped flint, tools which had been unknown to the later palaeolithic peoples of Europe. Both round-headed and long-headed individuals (the latter descended from the tall Cro-Magnon race of the West) were to be found among the Dobbertin population. Kossinna regards them as the ancestors at once of the Indo-Germans (Aryans) and the Finns, supposing that they spoke an agglutinative tongue from which Indo-European and Finno-Ugrian were subsequently evolved.

As the centuries passed the North Sea coasts sank and the salt water flowed into the old lakes, thus forming the Baltic. In this phase, beginning according to Kossinna about 6000 B.C., the climate was warmer than to-day; Europe in fact then enjoyed a "climate optimum" and the waters of the new sea swarmed with fish. The new and improved conditions entailed adjustments of habit on the part of the old fisher folk. The more conservative element, mainly brachycephalic, would have kept to the fresh water and perpetuated with but slight modification the bone industry of Dobbertin-Maglemose. These bone-users, some of whom now began to spread eastward, are henceforth termed by Kossinna "pre-Finns" (*Vorfinnen*), the ancestors of the Finns. The more adaptable section of the inhabitants of Scandinavia, mainly dolichocephalic, took advantage of the abundant prey afforded by the warm salt water and became a coastal population. These, called by Kossinna the *Ellerbek* folk, created the well-known culture of the Danish kitchen-middens in which it is especially the flint industry which is developed. Eventually they would have invented pottery, domesticated some of the local animals and begun to cultivate wild grains. They sent out colonists to Britain and North France who brought thither the flint-using civilization called Campignian. Others had gone eastward to Lithuania, Poland and Volhynia; Kossinna thinks they even reached Mesopotamia as Sumerians, so that the supposed affinity between Sumerian and Indo-European speech would be happily explained. None of these peoples were yet Indo-Europeans,—Kossinna calls them pre-Indo-Germans (*Vorindogermanen*)—but they were on the way to becoming Aryan.

And those who stayed at home in Scandinavia actually did make that advance at the same time as they began to polish their flint axes. Quite what the culture of the undivided Indo-Germans was like we are not told. They had embarked upon food production, garden culture and stock-breeding, about 4500 B.C. (on Kossinna's chronology), and they could polish flint, but they possessed no metals. However, the period of Indo-German unity did not last long. About 4000 B.C., "that cleft which was to divide the *satem* from the *centum* languages sundered the population." Those who were to pronounce *k* as *s* went southward through Silesia and created the Danubian civilization which we have learnt to know in Chapter VI. These were now South Indo-Germans. In Hungary they discovered and began to exploit the local copper ores, casting among other things the curious battle-axes of Fig. 27, 3, which they eventually introduced to the Caucasus. Moreover the whole culture with painted pottery in Transylvania and the Ukraine is attributed to the eastward expansion of the South Indo-German Danubians. But they were forced to surrender their territories to the advancing North Indo-Germans; the *satem* people were driven from the Ukraine to enter Greece as Thracians, even founding the Minoan civilization, while others were swept eastward to carry the art of vase-painting to Anau and Susa.

Meanwhile the section left at home on the coasts of the Baltic and the North Sea, now termed North Indo-Germans (*Nordindogermanen*), learned the art of building dolmens transmitted to Scandinavia via Ireland from Spain. And then began a period of rapid progress in the arts and of conquering expansion in all directions. The North Indo-Germans, now warriors, pastoralists, cultivators and navigators, sent out wave upon wave of warlike colonists.

Before considering these movements, which Professor Kossinna has traced in considerable detail in his earlier works, but has not yet reached in his latest monograph, we must pause to examine certain points in the foregoing theory. The account of the evolution of the early food-gathering populations may for the moment be accepted subject to the following reservations: the origin of the Maglemose culture must be sought in the south or south-east rather than the west[1]; the geological dates adopted by our author are far too high; the reference to the Sumerians is a baseless speculation and the spontaneous rise of agriculture on the shores of the Baltic seems unlikely. But no one who has read the previous pages of this book will agree to the view that the Danubians'

[1] *W.P.Z.*, xii, pp. 5–6

culture was derived from Scandinavia. As a matter of fact the German professor treats this as an axiom for which he offers no proof. "Up to date," he writes, "we have not succeeded in establishing by archaeological means a link between the North Indo-German and the South Indo-German cultures in such a way that the separation of the first from the second can be traced and a bond of union between them demonstrated. That is a painful gap in our insight into the archaeological material."[1]

We have further shown that the painted pottery of the Ukraine cannot simply be derived from that of the Danube valley. The idea of the foundation of the Minoan civilization by barbarians from Thrace is a fantasy which needs no refutation in English-speaking countries and even with Kossinna's inflated dates for the North it is a chronological absurdity to derive the painted wares of Elam and Mesopotamia from those of South-east Europe. All this means that Kossinna's South Indo-Germans must at once be eliminated from the Germanist scheme. And with all due deference I would submit that they are not really needed there at all. Still other Germanists besides the Berlin professor hold similar views. Schliz made even the earliest Danubians Nordics; Schmidt, Schuchhardt and Wilke derive the painted pottery of South-east Europe from the unpainted Danubian and call its makers Thracians.

Nevertheless, though we must abandon the identification of the Danubians with South Indo-Germans and the theory of their Scandinavian origin, there remain plenty of certain or at least plausible migrations from the North to which the spread of *satem* as well as *centum* languages can, if necessary, be attributed. In dealing with his North Indo-Germans, Professor Kossinna is in fact on much surer ground. The invading bands which he traces may be, if not North Indo-Germans, at least Aryans. And in respect of these wanderings there is a much larger measure of agreement; for the regions further removed from Germany the researches of other investigators, notably Åberg, Kozłowski, Menghin and Tallgren can be invoked. These authorities, though diverging from Kossinna on points of detail, are nevertheless animated by the same general conception. In what follows, I shall therefore attempt to give a synthetic picture of the conclusions

[1] 1921, p. 75.

of the Germanist school rather than to restrict my exposition to one single thesis.

We do find then on the shores of the Baltic and of the North Sea, this rude kitchen-midden population living in Denmark, Sweden and Norway, and contemporaneous with them remnants of the pure Maglemose-Dobbertin stock clinging to the fresh-water lakes. Even if it be admitted that the men of the kitchen-middens had made the first steps in agriculture and the domestication of animals, their life was barbarous and precarious. And for the most part they kept to the coast. A few scattered families may have wandered southward, but on the whole the primaeval forest formed an impenetrable barrier to the south hemming in the Baltic world. To this isolated circle came visitors from the sea, navigators from the south-west, seeking perhaps the source of amber. As Professor Kossinna remarks, the new arrivals need not have been very numerous, but yet they may have appeared to the rude fishers of the kitchen-middens as culture-heroes. They may even have established on those bleak coasts dynasties claiming divine honours and descent from the Sun-God, as Mr. Perry would have us believe, though through inter-marriage with the natives they would soon have been assimilated to the local population. In any case it is certain that the mariners from the West introduced to Scandinavia the cult of the dead and the megalithic funerary architecture associated therewith—first simple dolmens and then more pretentious structures termed passage graves. To the same people should in my opinion be ascribed the sudden improvement in the industrial arts and the beginnings of regular agriculture (garden-culture) and cattle-breeding.

Equipped with this new spiritual and material apparatus, the population began to force its way inland, obliged to find fresh tracts for tillage and grazing. Before 3000 B.C., on Kossinna's high chronology, pioneers spread westward along the North Sea coasts as far as the Zuyder Zee, building passage graves all over the heath-lands, and, urged by the poverty of the soil, pressed ever further southwards.[1] Eastwards more adventurous bands, driving their flocks before them and pursuing the game, followed the Vistula water-way, some to Galicia, some turning westward and reaching Silesia through Poznania.[2] This band, however, did not build megalithic graves but laid their dead to rest individually in small stone cists. On the Upper Oder they

[1] *Dawn*, pp. 205–210. [2] Ibid., pp. 228–235.

found Danubian peasants. Sometimes they exterminated these and settled on their lands—at Nosswitz in Lower Silesia a "Nordic" village of rectangular houses was built over the ruins of a Danubian settlement. Elsewhere they mixed with the Danubians; in the great cemetery of Jordansmühl (Upper Silesia) Nordic and Danubian II graves lay side by side. But the Danubians with whom the migrants here mingled, were not the simple peasants of the first period, but the more advanced people whom we have discovered at Lengyel in Hungary with Nordic skulls (p. 150). Should we invert the account given of the genesis of that group, seek its origin in Silesia, and derive thence the Nordic element found in Hungary? That would be a bold step, and Silesian, Bohemian and Moravian archaeologists are not prepared to make the experiment.

In any case the same Nordic stream which had reached Silesia pursued its course westward towards the Elbe and the Saale, there to meet other currents; for all this time a steady expansion had been taking place southward towards Central Germany. Its monuments are the megalith graves and Nordic pottery which cover all North Germany and penetrate ever further south along the Saale and Elbe highways. But it must be remembered that the Nordic cultures in Germany are very far from homogeneous. Some peoples buried their dead collectively in megalithic tombs, others in regular cemeteries of separate graves; the variety of the pottery is bewildering; both long and short-headed skulls are met. We get the impression of a tumultuous flood of rudimentary clans or tribal groups in continuous interrelation. Often they were at war, for the multitude of stone weapons is innumerable. But regular trade relations subsisted between the various groups illustrated by the diffusion of amber and other commodities.

But here we must digress to examine a very puzzling phenomenon at the heart of the turmoil, in Scandinavia itself. For there, especially in Denmark, no less than three distinct civilizations are to be distinguished. First come the descendants of the Dobbertin (or Ellerbek) folk in Norway and inner Sweden, who were as yet unaffected by the civilization of the megalith-builders and lived on as food-gatherers, using bone very largely or translating bone implements into slate. Then on the coasts and spreading, as we have seen, southwards were the megalith-builders. The third group was very different to either of the foregoing.

In constrast to the megalith-builders whose sepulchres were collective tombs where the members of a family or tribe were buried together for generations or to the food-gatherers who do not seem to have observed any regular burial rites at all, the third people were interred in separate graves, one for each person, lined with stones and surmounted by a barrow.[1] The oldest of these graves in Jutland are contemporary with the dolmens on the coast and often contain a similar furniture. But in the succeeding period the separate graves constitute a quite isolated group. Not only their form but also their furniture—pottery ornamented with cord impressions, spheroid maceheads, peculiar types of stone battle-axes, and special kinds of ornaments—is utterly different to that of the contemporary collective tombs, the passage graves. Moreover, these separate graves occupy the whole of inner Jutland to the exclusion of megalithic tombs. There is no doubt that they belong to a distinct and war-like population—we may call them battle-axe folk—who had checked the expansion of the megalith-builders in this direction and who largely lived on tribute exacted from their coastal neighbours.

Who are these people? Kossinna says that they are descendants of the "pre-Finnish" Dobbertin stock in course of Indo-Germanization. Was this predatory folk generated and organized by discontented scions of the ruling houses of the "Archaic (megalithic) Civilization" on the coasts in the way that Perry [2] describes—that might correspond to Kossinna's "Indogermanization"? Or finally were the warriors invaders come from more continental regions as Sophus Müller, Knut Stjerna and C. A. Nordmann contend? Similar people are certainly to be met in Thuringia, South Sweden and Finland. The Thuringian barrows cluster all along the hills, dominating the Saale salt deposits and the Elbe amber trade route as if their builders had been taking toll on Central German commerce, just as their kinsmen in Jutland did on the coastal traffic. And the Thuringian barrows cover graves of similar form and furnished with similar objects [3]—cord-ornamented vases, stone battle-axes and spheriod mace-heads—to those of Jutland. A less strict parallelism links the battle-axe graves of Sweden and Finland to those of Jutland and Thuringia and to one another. We have

[1] *Dawn*, pp. 206, 209–11.
[2] *Growth of Civilization*, caps. vii and viii.
[3] *Dawn*, figs. 116, 117, 100.

used the word parallelism advisedly, for it is extremely difficult to derive any one group directly from any other.

Before concluding this digression, let us state that the origin of the battle-axe folk is the crucial question for the Germanists; not only are the skulls, from the Thuringian barrows at least, typically Nordic but it is to the battle-axe folk that several cultures which can be identified as the work of Aryans are directly linked. The round-barrow men of Britain and the proto-Celts buried in the Bronze Age barrows of Bavaria are the direct descendants of the battle-axe folk from Thuringia who lie beneath the Stone Age barrows of the Rhineland [1]; even in Scandinavia and North Germany it is the civilization of the separate graves which eventually becomes dominant, absorbing the megalith culture and that of the fisher-folk so that they leave no trace behind. For the moment, however, we shall accept Kossinna's doctrine [2] of the local origin of this culture in Jutland, and treat the battle-axe cultures of Thuringia, Sweden and Finland as derived therefrom. With this assumption we can trace more closely the expansion of the several waves of Nordic tribes.

The tumultuous bands who pressed southward though Central Germany did not first penetrate south of the heights round Magdeburg; further south lived the Danubian peasants. But there came a time when the rough Nordics entered the Danubian province. At first they advanced by slow infiltration, and individual families or groups were admitted to membership of Danubian communities. The presence of genuine Nordics among the Danubians when they had at a relatively late date colonized the Rhineland (p. 141) from Thuringia could be thus explained. But ultimately the barbarians from the North fell upon the peasant communities and finally overlaid them—we have already described the process. And even before the central wedge had crossed the Danubian frontier on the Elbe and Saale, the more easterly band of invaders, whom we left in Silesia a few pages back, had been advancing southward into Moravia and westward to the Elbe. Some of these same migrants must have reached the eastern slopes of the Alps; for stone battle-axes like those found in the Austrian lake-dwellings and the land stations of

[1] I hold that the barrows of the "Tumulus Bronze Age" in the South German highlands belong to descendants of these "neolithic" people, some of whose interments must be contemporary with Early Bronze Age graves in the valleys.

[2] This view is ably sustained by Menghin in Hoernes, *Urgeschichte*, pp. 738 and 767.

Bavaria have been met in the cist graves of Galicia (Fig. 27, 5, cf. p. 156 above). The other branch of the same band had, as we saw, proceeded eastwards into Galicia and beyond.

A second wave[1] of peoples penetrated still further into South-east Europe. Starting from Denmark, as Kossinna now thinks, or from Central Germany as he held in 1910 and Åberg subsequently argued, they spread to the Dniestr valley and the Ukraine and possibly even reached the Caucasus; the monuments of this invasion are in the first place globular amphorae,[2] and it is certain that such vessels from Eastern Galicia and the Ukraine are identical in form and decoration with those found in Central Germany. To the same migrant hordes Kossinna attributes the erection of the megalithic tombs found between the Oder and the Vistula. In 1902 he ascribed the "dolmens" of the Black Sea coasts to a like body of Nordic invaders. Though this suggestion has not been repeated in subsequent articles, it is certain that some sort of connection between the Caucasian "dolmens" and part of the Nordic megalithic culture exists; two peculiar double dolmens on the Kuban in the Caucasus are identical in form with one at Baalberg in the Saale valley and contain very similar vases.[3] Though the Caucasian tombs were furnished in addition with metal objects which show Mesopotamian influence, it is not inconceivable that they were built to the order of some Nordic chief from Germany. And later on similar curious megalithic tombs were built south of the Caucasus on the Caspian coasts. Here if we liked we might see the vanguard of Nordic hosts advancing on Iran to become Indo-Iranians.

Even wider was the range of the nomadic warriors who buried their dead in separate graves under barrows with cord-ornamented vases. Westward they spread from Thuringia to the Rhinelands where they stood as overlords among the hill population. Mixed with Prospectors they set out to invade Britain as the round-barrow men (page 162), while others remaining behind in the Rhineland took to burning their dead (page 145), invaded Switzerland with their battle-axes and corded vases (page 157) and in the highlands of South Germany formed the nucleus of the Bronze Age barrow-builders to whom the north-western Hallstatt civilization may later on be ascribed.[4] At the same time

[1] The three waves of invasion are described with maps in *Mannus*, i–ii.
[2] *Dawn*, fig. 114.
[3] *Dawn*, figs. 63, 1 and 62.
[4] The continuity between the (Middle) Bronze Age and Early Iron Age

a similar people had reached the coasts of the North Sea in the Stone Age and there had superseded the megalith culture of Holland by their own. To the south we have already seen how the second wave of Nordics as conquerors of the Danubian peasants introduced cord-ornamented pottery into Moravia and Bohemia while the ceramic evidence revealed similar intruders entering Hungary and Transylvania from the north and the south-east (p. 151).

Eastward the battle-axe folk travelled even further. In South Russia, where they represent the third wave of the Indo-Germanic exodus, Kossinna has traced them to the Black Sea and the Don; the barrows there contain stone battle-axes and vases ornamented with cord-impression just as in Jutland or Thuringia, and the corpses belonged to true Nordics. The flint celts and stone battle-axes which appear as strange intruders among the vase-painters of Bulgaria might be assigned to a branch of the same stream, and Åberg[1] has derived the ceremonial weapons of the lords of Troy from the Danish axes through Russian types (it is certain that axes of distinctively East European type closely akin to those from Silesia were found at that city).

Yet another migration has been traced by Professor Tallgren,[2] who regards the wanderers as Indo-Germans, though Kossinna calls them "Finns". In the valley of the Upper Volga are several cemeteries of separate graves lined with stones much like those of Jutland; they contain flint celts, necklaces of teeth, vases not very different from those of the Swedish battle-axe graves and battle-axes which Tallgren now regards as descended from the Danish. This civilization is called the "Fatyanovo culture". According to Professor Tallgren it was due to the advance of warlike Nordic tribes from Scandinavia, perhaps through East Prussia or Finland where battle-axe cultures are also known, to Central Russia. Locally this author finds no continuation for the Fatyanovo culture, but in the Caucasus, in the brilliant copper age of the Kuban, it would be propagated further. No doubt the wealth of the Kuban graves comes from Mesopotamia, but as the spoils of Mesopotamia only. "As in the great migrations of the first centuries of our era, Nordic barbarians

populations here is proved by the pottery (Behrens, *Bronzezeit*, p. 276). But the Central Hallstatt culture belonging partly to Illyrians (Menghin, op. cit., p. 841) pushing up from the south-east produced mixtures and displacements in the population (see Keller and Reinerth, *Thurgau*, p. 76).

[1] *Das nordische Kulturgebiet.*

[2] "L'âge du cuivre en Russie centrale" in *S.M.Y.A.*, xxxii; cf. *Dawn*, pp. 224–6, figs. 108–9.

had occupied the seats of the Oriental kings and buried in their tombs the plundered treasures."[1] Here then we would have a stream of Nordic Aryans, crossed, if you will, with another current coming across the steppes, but well on their way to the Ancient East where they would emerge as Mitanni kings, Hittites, Persians or Hindus.[2]

Finally in Scandinavia and North Germany itself the civilization of the Bronze Age, which must belong to Teutons, is a direct continuation of that of the Stone Age, fertilized indeed by trade with Britain, Bohemia, Hungary and Italy, but developing without a break in population or culture.

Here then we have in outline a picture of the expansion of Nordic civilization and its ruling race from Scandinavia and Germany. The events of prehistory faithfully anticipate the great migrations of the first centuries of our era. But these migrants who gravitated towards the centres of antique culture and conquered them in the late Stone Age were not yet Germans but Indo-Germans, Aryans. Where we have followed their wanderings, we have found them eventually emerging as Celts, Italici, Hellenes, and Indo-Iranians. As for the Finns, whose linguistic relation to the Aryans is an essential part of the theory of Penka and Kossinna, they are easily identified. From Norway to the Urals we know a rude neolithic culture characterized by bone implements, or imitations thereof in slate, round-bottomed pots, and a naturalistic sculpture.[3] These artifacts are held to betray the kinship of their makers with the Dobbertin-Maglemose folk and consequently their western origin. But confined to the forests and swamps of the frozen North this population remained long in the food-gathering stage of culture, despite a rudimentary system of barter and contact with the battle-axe folk. These rude hunters may well be the ancestors of the Lapps and Finns, though the latter only developed into a specific people at a much later date in the Ural region. Thus the relation of Finns to Aryans would be satisfactorily explained. We should further be in a position to answer an objection raised by de Michelis and others to the hypothesis of a North (or East) European cradle: "Why," they ask, "granting the peculiar virtues of the Nordic race and its

[1] Tallgren, in *Finskt Museum*, 1924, p. 25.

[2] Tallgren, adopting a conservative estimate, dates the Kuban copper age between 2000 and 1500 B.C., *Studia Orientalia Fennica*, 1925, pp. 340 ff.

[3] *Dawn*, pp. 219–222.

Aryan speech, were the Finns not Aryanized?" We might now reply: Because they were relegated to the inhospitable forests and swamps, and such sub-arctic regions and their savage denizens exercised no attraction on the domineering Aryans who preferred enjoying the fruits of others' labour to themselves opening up the pathless woodland.

As thus presented the Germanist doctrine is the most comprehensive and consistent synthesis of Indo-European peoples that has ever been offered. It is the only doctrine the extant expositions of which can pretend to combine the results of recent archaeological research with the data of philology. At the same time it is one of the fairest and certainly the most economical account of the development of a peculiarly European civilization yet propounded. Indeed, if it can prove its validity in the realm of archaeology and ethnology, it will probably have to rank as an accurate solution of the Aryan question. In these respects, however, it is to-day not quite unassailable. Some of the objections are of a highly technical nature; for a discussion of these I refer the reader to my *Dawn of European Civilization.* Here I shall limit myself to a few more general points.

Firstly in the sphere of ethnology, the bases of the theory are not so stable as might be wished. The skulls on which Kossinna relies to prove the Nordic character of his Maglemose-Dobbertin folk are by no means certainly dated; in any case the Nordic race can scarcely be derived from the western Cro-Magnon stock, but had eastern or Central European antecedents.[1] It can nevertheless be regarded as generally probable that a sort of proto-Nordic element was present in the North in the days of the Maglemose culture and of the later kitchen-middens, as it had been in the last phase of the Old Stone Age in South Germany.[2] On the other hand the bodies interred in the early dolmens, as Kossinna himself points out, belonged according to Karl Fürst to individuals who, although dolichocephalic, were short of stature, i.e., to members of that same Eurafrican race which built the other dolmens in Western Europe and the long barrows in Britain.

In the second place the civilization of Denmark and Scandinavia at the epoch of the early dolmens is not wholly explicable either as the product of Western and South-western inspiration or as the result of a local and spontaneous evolution. The pottery,

[1] *W.P.Z.*, xii, pp. 8–9.
[2] E.g. the skull from Ober-Cassel (Magdalenian), Keith, *Antiquity*, i, p. 108.

especially curious little flasks with a clay ring or collar round the neck (collared flasks),[1] and battle-axes and other objects found in the dolmens have no prototypes in the realm of the megalithic culture further south and west. At the same time the subsequent evolution of arts and industries in Sweden, Denmark and North Germany, was far more rapid and brilliant than in the megalithic provinces of France or Spain. But if the objects from the Scandinavian dolmens cannot be regarded as introduced and inspired from the West, they can still less be regarded as the spontaneous inventions of the local native authors of the kitchen-midden culture; the continuation of the latter is to be found in the "dwelling places" of South Sweden belonging to a backward race of food-gatherers. Their pots, for instance, do carry on the tradition of the kitchen-middens, as those from dolmens and separate graves do not.

Now Sophus Müller has suggested that the curious vases, such as the collared flasks, found in the Danish dolmens which are so hard to explain came in fact from the South-east; in Silesia and Galicia they are not uncommon and are found in separate graves sometimes accompanied by perforated stone battle-axes of a type found in Danish dolmens.[2] Might we not regard such objects from dolmens as borrowed from the people who buried their dead in separate graves?

Thirdly the culture of the separate graves in Denmark raises a very puzzling problem. Kossinna as we saw attributes these interments to descendants of the aboriginal "pre-Finnish" population surviving from times anterior to the oldest dolmens. Scandinavian archaeologists are not altogether averse to such an explanation as far as the oldest separate graves, contemporary with the dolmens, are concerned. But when we come down to the epoch of the passage graves, the majority of competent authorities, Sophus Müller, Knut Stjerna and C. A. Nordman, look to an invasion to account for the battle-axe folk whose separate graves occupy the interior of Jutland to the exclusion of the collective megalithic tombs. In the case of similar battle-axe graves in Finland there is in fact no possibility of doubt that they were dug by intruders and the same may well be true of the corresponding interments in Sweden. All this is very hard to reconcile with Kossinna's hypothesis, for the same "pre-Finnish"

[1] *Dawn*, pp. 205 ff. and fig. 98.
[2] Ibid., fig. 112.

population to which he attributes the Danish burials was available also in Finland and Sweden, where it did not develop into a battle-axe folk but was overcome thereby.

If then we must admit invaders even in Denmark, whither should we look for their homeland and starting point? Some might point to Thuringia. But the Thuringian barrows with cord-ornamented pottery seem rather parallel than prior to the Danish separate graves. Nor is a local origin of the Thuringian culture really discoverable, though Götze, Schliz and Schuchhardt think differently; there too it looks as if we had to do with intruders, probably a band of the same stock as had invaded Denmark. But if this be true it is fatal not only to a large part of Kossinna's special theory, but to the whole idea of a Scandinavian origin for the Aryans; for it is with the separate grave folk, wielders of battle-axes, and not the megalith-builders that the European cultures wherein we would detect Aryans, are to be connected. That is true not only of the proto-Celts in South-west Germany, but even of the Teutons themselves, since it was the separate grave culture that eventually obtained the mastery in Scandinavia by the beginning of the Bronze Age.

Finally there are very grave chronological obstacles in the way of regarding the battle-axes of Troy and Hungary, which seemed to us the most "Aryan" elements there, as descended from the Danish or Thuringian. With Kossinna's inflated chronology indeed there would be no contradiction in so doing, but that chronology rests on no sure foundations and ultimately leads to results highly disadvantageous to the Nordic peoples whom he wishes to exalt. On the other hand the dates given by relations through the Danube valley and through Britain with the Aegean where alone an absolute chronology is available before 600 B.C., would place the Danish dolmens not much before 2600 B.C., and the earliest passage graves with which the first expansion of the battle-axe cultures should coincide, about 2200 B.C. Plainly that will not allow the Nordics from Scandinavia to have reached the Troad before 2200 B.C.

No doubt these objections are not insuperable. They may all be eliminated as a result of further investigation, and in that case the Germanist theory would probably be acceptable. But in the interim we are at liberty to seek an explanation of the unintelligible phenomena outside Scandinavia and in so doing to look for an Aryan cradle that harmonizes better with the data of linguistic palaeontology. Indeed, no one can fail to be struck by the discrepancies between the picture of Aryan culture sketched in Chapter IV and that derived from a study of the antiquities of Scandinavia. We come then to the hypothesis first propounded, very cautiously, by Professor J. L. Myres, and developed more recently by Mr. Harold Peake. We propose in a word to invert all the eastern and south-eastern movements traced on the principles of Kossinna and Tallgren in the present chapter and, following a hint given by Sophus Müller, to derive the battle-axe folk of the North, who were so evidently Aryan, from South Russia.

CHAPTER VIII

THE ARYANS IN SOUTH RUSSIA

Having surveyed all other regions of Europe we turn to the South Russian steppes. The climate and physiographical features thereof, as Otto Schrader so convincingly argued, correspond admirably to the characters of the Aryan cradle as deduced by linguistic palaeontology. And the earliest connected remains of post-glacial man there likewise reveal a culture [1] which harmonizes to a remarkable degree with the proto-Aryan culture described by the philologists. The remains in question are derived almost exclusively from graves containing contracted skeletons covered with red ochre (ochre-graves) and surmounted by a mound or *kurgan.* The people here interred were generally tall, dolichocephalic, orthognathic and leptorhine, in a word Nordics. There was, however, at least a small minority of brachycephals present in the population.

The material from the oldest *kurgans* is poor and rude, yet it is relatively uniform over the whole area from the Caspian to the Dniepr. This cultural uniformity would perhaps allow us to infer the currency also of a single language in the sense explained on p. 11. Again the strict observance of the same peculiar burial rites over the whole area might betoken a community of religious ideas among all the *kurgan*-builders which would also have been expressed in the recognition of one or more common deities. It would be tempting to call that common language Indo-European and the common deity *Dyeus,* since the furniture of the graves reveals a culture extraordinarily similar to that described in Chapter IV.

In the first place these Nordics of the steppe were pastoralists; since the bones of animals are found in the *kurgans.* The remains include not only sheep and cattle but also the bones of that peculiarly Aryan quadruped the horse. Though the exact race does not seem to have been determined, it may be assumed

[1] The evidence on which our knowledge of this culture is based is summarized in *Dawn,* chap. x. Add now Tallgren's articles in *S.O.F.*, 1924, and in *Götze-Festschrift,* 1925.

from the general nature of the country that the animal in question was related either to the swift desert horse as found by the Americans in Transcaspia (p. 109), or the steppe horse of Przybalski[1] and

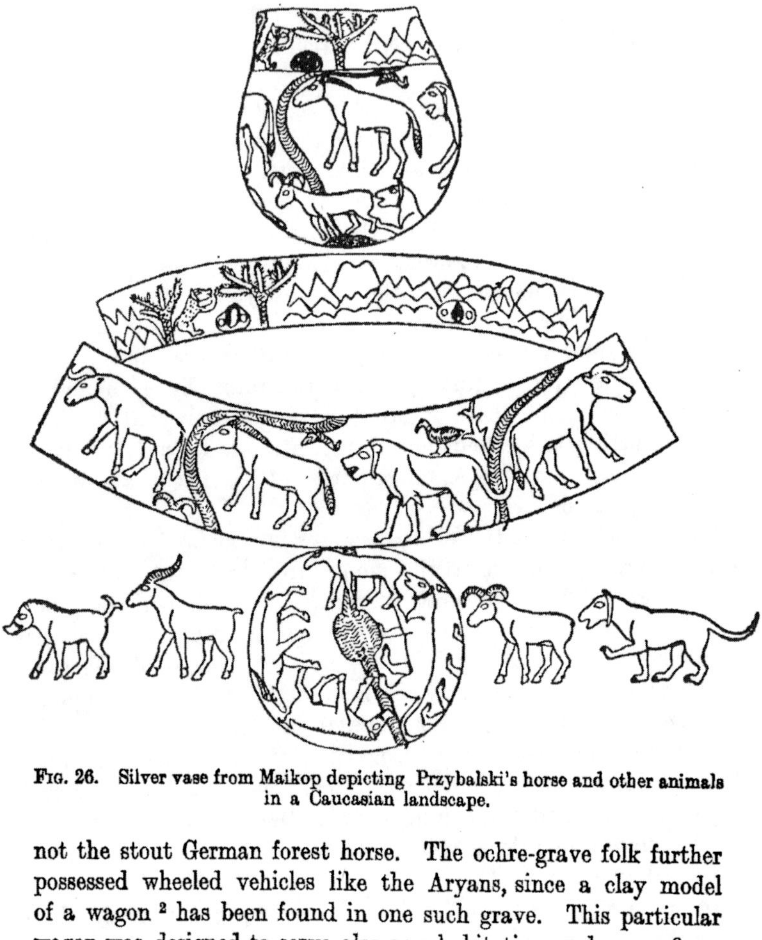

FIG. 26. Silver vase from Maikop depicting Przybalski's horse and other animals in a Caucasian landscape.

not the stout German forest horse. The ochre-grave folk further possessed wheeled vehicles like the Aryans, since a clay model of a wagon[2] has been found in one such grave. This particular wagon was designed to serve also as a habitation, and so confirms Peake's inference from the poverty of the graves that our people were partly nomadic like the Scythians and Getae, described by classical authors. However a plurality of interments at different levels in the same barrow betokens a continued occupation

[1] The latter is represented on the silver vase from Maikop, Fig. 26.
[2] *Dawn*, fig. 64.

of certain districts for a longer or shorter time. Moreover, if not from the first and everywhere, the prehistoric inhabitants of South Russia did practise a little agriculture; for grain has been found in some *kurgans*. In fact, at a mature phase of their development, some of them began to settle down in regular villages in the more fertile valleys and on the coasts.

Again the Nordics were here in a chalcolithic phase of culture. In the oldest graves indeed implements and weapons of stone, flint and bone predominate, but almost everywhere small articles of pure copper (p. 85) are to be found, many of them obviously importations. Silver is also fairly widely distributed; and indeed is more common here than anywhere else in Europe at the same epoch; gold is met only in the Kuban valley. Of course the metal objects are in an immense minority, save in the Kuban region. Among the tools besides flat celts of flint or copper, bone prickers and quadrangular copper awls deserve especial mention. The South Russian armoury corresponds closely to that deduced for the Aryans. Perforated axes of stone or copper (*peleku*) are particularly common and some of them are demonstrably imported from Mesopotamia (p. 87). Flint arrow-heads indicate an acquaintance with the bow as clearly as our equations on p. 92. Flint and copper points are widely distributed and these—especially the copper blades—could equally have been attached to the end of a long pole for use as pike-heads, or fitted with a short handle to form daggers, reminding us of the change of meaning between Sans. *śaru* "spear" and Goth. *hairus* "sword" (p. 85).

This concordance between the linguistic and the archaeological data is itself very striking, but we might go considerably further. Philology suggests contact between the undivided Aryans and the Sumero-Akkadians. Just so the industry of the steppe graves bears upon it the unmistakable imprint of Mesopotamian civilization in the creation of which the Sumerians played the leading rôle. The earliest metal types from South-east Russia, concave chisels, spear-heads, perforated axes, belong essentially to the Asiatic as distinct from the Egyptian, Minoan or West European series. That implies that metal reached the steppe from the Mesopotamian region and we know that one Indo-European word for copper is derived from the Sumerian (p. 87). Again the Aryan word for axe was borrowed from the same direction. Not only are the copper axes from South Russia obviously allied to types in use by the Sumerians from the IVth millennium B.C., but one grave at Maikop on the Kuban contained a battle-axe shaped like a hoe with the blade at right angles to the shaft (Fig. 27, 2). This weapon was unquestionably an import from Mesopotamia, since the type is not met elsewhere outside the Tigris-Euphrates valleys, where it was in use from about 3500 to 1100 B.C. Moreover clay figurines of naked women are found, although extremely rarely, in ochre-graves; these bear a distinct likeness to models of the goddess Ishtar found at Assur and elsewhere in Mesopotamia. Now it has been suggested that this divine name is concealed in the Indo-European word for "star" *ęster* (p. 87) and the ideogram for Ishtar in Babylonian was precisely a star.[1] Thus the connections with Mesopotamia postulated by philology for the Indo-Europeans are proved to have been a reality among the early nomads of South Russia.

On the other hand, if we agree with Pokorny that the word *ayos* "copper" comes from *Alasya* and indicates intercourse between the Aryans and the Aegean peoples, traces of such connection are not wanting in South Russia. At least the later ochre-graves near the mouth of the Don imitate in shape the "pit-caves" (a sort of chamber tomb) in use in the Aegean by the IIIrd millennium and contain ornaments, such as phallic beads, that prove trade with the Cyclades. Finally the assumed connection between the Aryans and the Finno-Ugrian peoples would be as easily explained on the South Russian hypothesis as on the Scandinavian, for the same rude hunting folk who made the Swedish "dwelling place" culture, were spread far and wide through Central Russia, and there is plenty of evidence for contact between the two areas. On the one hand the barbaric pottery characteristic of the northern forest belt extends southwards to the edge of the steppe, on the other we find the same pottery in Central Russia, associated with daggers, copper battle-axes and idols of Babylonian type[2] that can only have come from the south, across the steppes.

Can we then call these "neolithic" people of the steppes Aryans without qualification; or were they just a branch of that stock as the Germanists contend? Professor Myres, Professor Haddon and Mr. Peake all incline to the former hypothesis without,

[1] Delaporte, p. 140.
[2] *Dawn*, p. 221 and fig. 106.

however, offering any body of detailed evidence in refutation of the contrary view. Their thesis plainly implies in the first place the existence of a pre-neolithic population in South-east Russia, and secondly that this population, having acquired or elaborated the neolithic civilization described in the preceding paragraphs, sent out bodies of emigrants to carry that culture to the rest of Europe.

The first point is capable of proof. Mr. Peake has suggested that the ochre-grave folk were descended from the Solutreans who had hunted the horse in western Europe in the Old Stone Age. Now the Solutrean phase of the Old Stone Age is in fact well represented in the Ukraine, as well as in the Caucasus. And although nothing exactly parallel to the later phase represented in France by Magdalenian industry has yet been found in this direction, evidence for a continuous occupation of the southern plain of Europe is rapidly accumulating. Not only is that presupposed in the migration from the east postulated by some authors to account for the establishment of the Maglemose culture on the Baltic; there is even less ambiguous evidence for a drift of people from the same quarter at a still earlier date, corresponding roughly to the last phase of the reindeer age in France. The earliest remains of human handiwork yet discovered in Scandinavian lands include pigmy flints arrow-heads.[1] In form and technique these are quite foreign to the microlithic industries of Western Europe, but they are characteristic of the earliest microlithic culture on the sand-dunes of Little Poland.[2] This industry therefore was introduced into Scandinavia from the east in pre-Maglemose times. Moving further west, it just reached the coasts of Yorkshire[3] contemporaneously with the Maglemose culture. So there must have been an overflow of people from the south-eastern plain in the wake of the last glaciers. There must consequently have been people in Poland and *a fortiori* in the more habitable regions of South Russia at an earlier date. And, as a matter of fact, other pigmy flints have been discovered on the banks of the Desna, Dniepr and Don, in the Crimea and on the Kirghiz Steppe.[4] The latter are parallel to the Tardenoisian

[1] At Lingby in Jutland and at other sites in Denmark and Norway, *W.P.Z.*, xii, pp. 1 ff., fig. 1, 1–5.

[2] Kozłowski's Chwalibogowice type, *Dawn*, p. 11, fig. 7.

[3] Recently discovered by Professor Kozłowski during a visit to England in 1925 among the flints from Holderness.

[4] *Russ. Antrop. Journal*, 1924, pp. 211 ff. (with English resumé).

industry of France (assigned to an epoch intermediate between the Old Stone Age and the New) but in view of analogues in Mesopotamia, India and even Mongolia may well belong to an independent group.

Hence the presence of pre-neolithic men on the eastern portion of the European plain is demonstrated and it is clear that they were at this time drifting westward. They might be Haddon's proto-Nordics ;[1] the Nordic skull from Ober-Cassel and the Nordic elements in Maglemose and the kitchen-middens would mark outposts of their post-glacial advance westward. We would in fact have a sparse population of proto-Nordic hunters disseminated unevenly from the Black Sea to the Baltic by early post-glacial times. They would not yet be Aryans, but we might suppose that those who settled in the north became the ancestors of the Finns. The view here advocated would be that another section of this proto-Nordic stock, concentrated on the Pontic steppe, developed there the neolithic civilization of the ochre-graves and then diffused it to Central Europe. The Germanists on the contrary contend that the kernel of the ochre-grave culture was brought fully fledged from Scandinavia. It is possible to give certain arguments in favour of our view.

We have seen in the preceding chapters that the characteristic attribute and symbol of the Nordic cultures which we now recognize as Aryan was the perforated battle-axe. Now the genesis of this very peculiar weapon can be explained in South Russia better than anywhere else. Such weapons are far more unusual than might be thought. Very few peoples have hit upon the seemingly simple plan of putting the shaft of the axe through its head. The ancient Egyptians till Hellenistic times, the prehistoric inhabitants of Western Europe down to about 1000 B.C., the pre-Columbian Indians of America, the Pacific Islanders before the advent of Europeans and many other primitive peoples all used the clumsy device of tying the axe-head of stone or metal on to, or into the cleft of, a stick! On the other hand from the Alps to the Zagros proper perforated axes with a shaft-hole in the head were in use from the IIIrd millennium before our era. It would be natural to infer that this exceptional device, employed only in such a relatively limited area, was invented in one single centre and

[1] Ekholm, however (*Ymer*, 1924; *W.P.Z.*, loc. cit.), would connect the introduction of the Lingby culture with a brachycephalic race. He seems also to regard all proto-Nordics as already Aryan.

diffused thence. To-day it is reasonably certain that that centre was Mesopotamia. The recent English and American excavations at Ur and Kish have brought to light actual specimens and clay models of perforated copper axe-heads, dating from the IVth millennium B.C. Even on the Germanist chronology these are quite the earliest dated examples of such weapons.

Moreover, a good case can be made out for the belief that the idea, born in Mesopotamia, was transmitted to the rest of Europe from the North Caucasus precisely by our Nordics. Eminent Scandinavian archaeologists[1] have long recognized that the Northern stone battle-axes were imitations of a curious copper weapon with one blade parallel and one at right angles to the shaft, conventionally termed an axe-adze (Fig. 27, 3), citing well-known examples from Hungary. But this freakish implement itself requires explanation, and that cannot be found in Hungary, but only further east. The Sumerians by 3000 B.C., were using two types of copper battle-axe in one of which the blade is parallel to the shaft and in the other at right angles as in a hoe (Fig. 27, 1-2). The only intelligible explanation for the Hungarian axe-adze is to regard it as an amalgamation of the two Mesopotamian types. Now in Babylonia and Assyria this compound type is not found till about 1100 B.C., but there is a specimen from a "treasure" or tomb-group, dated by Sumerian gold vases included in it to the IIIrd millennium, found many years ago in a mound near Astrabad, south of the Caspian.[2] Moreover, there is another example from an ochre-grave at Maikop on the Kuban which also contained an axe of the peculiarly Mesopotamian hoe-like type. Somewhere in this corner of the world then the axe-adze might have been invented. Its translation into stone among peoples lacking copper ore would account for the Nordic weapons.[3]

As a matter of fact we would get, as will appear shortly, a very good distribution both for the copper prototypes and for the stone copies if we supposed that they radiated from a focus in South-east Russia. At the same time we should avoid the chronological difficulties presented by the Trojan axes if we assumed that they are the result of a parallel and contemporary evolution and not

[1] Montelius, in *AfA.*, 1899, and Knut Stjerna, *Före Hällkisttiden.*

[2] Rostovtseff, *J.Eg.A.*, vi, pp. 6 ff. Frankfort very properly points out that the Sumerian vases do not indicate the presence of a Sumerian *patesi* in North Iran, but are the loot brought home by some local chief, *Studies*, i, p. 85.

[3] Some of the stone battle-axes from ochre graves are quite obviously imitations of copper prototypes as Professor Tallgren has pointed out (*S.M.Y.A.*, xxv, p. 126).

descendants of the Scandinavian. Here then is strong inferential evidence for the belief that the battle-axe folk of North Europe came from the South-east and not vice versa.

Of course this is far from constituting proof. A typology is a two-edged weapon unless both ends of the series are safely dated.

A second possible argument for our view is to be found in the distribution of prehistoric equidae. We have seen (p. 109)

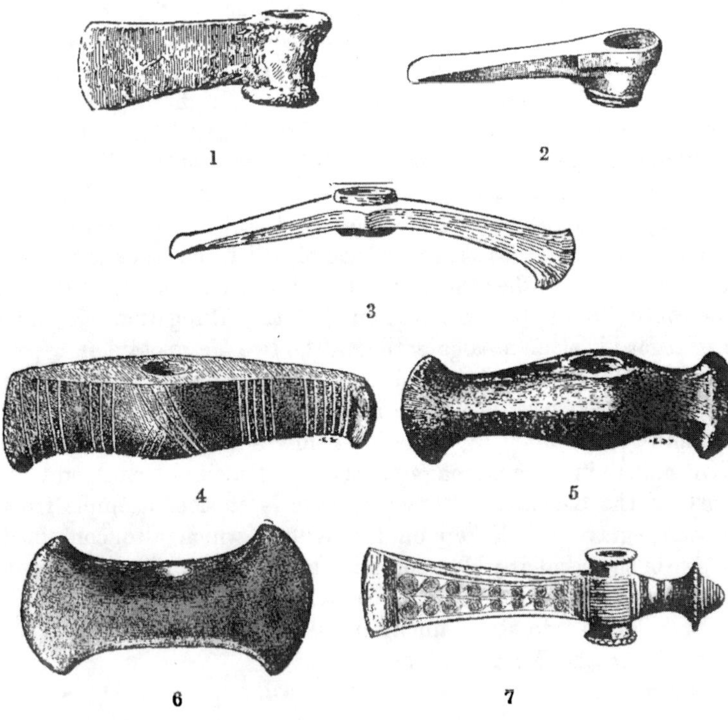

FIG. 27. The typology of the Battle-axe. 1–3, Copper prototypes: 1–2, Mesopotamia; 3, Caucasus and Hungary; 4–6, Stone copies: 4–5, Silesia; 6, Britain; 7, Bronze derivative: Scandinavia.

that the swift horse first appears tame in Transcaucasia and that this horse was the ancestor of the Bronze Age horses of Europe. Who were more likely to have introduced this animal to the western world than our nomadic people of the steppes? Mr. Peake indeed thinks that they were responsible for domesticating the beast their ancestors once had hunted. If it could be shown that the swift horse appeared in Europe simultaneously with the battle-axe cultures, we should have a really conclusive argument in favour of our view. At the moment unfortunately all that can be proved is that remains of the swift Asiatic horse and evidence for the domestication of equidae are only found in Central Europe after the spread of the battle-axe cultures. The material available is exiguous; to determine whether the animal be domesticated

FIG. 28. Palæolithic drawings of horse and mammoth from the cave of Combarelles, Dordogne (Magdalenian).

or not is peculiarly difficult, even the distinction between the Asiatic horse and the heavier variety native to the forests of northern Europe can only be recognized by a specialist.

Bearing this in mind we may say that the descendants of the Anau horse are first certainly recognizable in Swiss lake-dwellings of the Late Bronze Age (about 1000 B.C.).[1] It is, however, likely

[1] Duerst in Pumpelly, op. cit., ii, p. 429; for the chronology the finds from the station of Alpenquai, Zurich (*M.A.G.Z.*, 1924, p. 193), are decisive.

that the horses' bones found with stone battle-axes in a fortified settlement of the Copper Age near Hammerau in Bavaria[1] belong to the same breed. At the same time the earliest certain evidence for the domestication of horses—horn bits—does not take us back beyond the Middle Bronze Age in Central Europe (about 1500 B.C.); only the bit from Gross Czernosek on the Elbe in Bohemia may be rather earlier if the defective report on the excavations be accepted.[2] The precise coincidence desired is not therefore established. It remains possible that the Anau horse came to Europe with the vase-painters before the battle axe cultures, and was slowly diffused from Transylvania to Switzerland and Bavaria even though he was not associated with the other domestic animals introduced into Europe at the beginning of the New Stone Age. It is also possible that the native forest horse was independently domesticated in the North.

The indications furnished by the battle-axes and the horse cannot, therefore, be regarded as conclusive in view of the mass of evidence collected by the advocates of the Germanist thesis. We may, nevertheless, examine further the implications of our theory.

The Migrations of the Aryans

We should begin with a sparse population of pre-neolithic hunters strung out widely over the steppe. In South Russia we may at least say that the conditions would be favourable for their initiation into precisely those rudiments of neolithic culture that characterize the Aryans. To the east the vase-painters would have settled at Anau. The fertile black-earth tracts on the west were early occupied by similar agriculturalists. Both or either of these groups could have acted as masters to the nomads in the arts of food-production. South of the Caucasus and the Black Sea lay Mesopotamia where a great civilization had been flourishing from the end of the Vth millennium. There is no doubt whatsoever that that civilization did influence the people

[1] *Beiträge z. Anthrop. u. Urgeschichte Bayerns*, x, p. 192; xi, p. 308 ff. The horse from Misskogl near Lesskoun in Bohemia may be Early Bronze Age (*M.A.G.W.*, xx, p. 133). The bones from O. Besseny on the Danube belong to a period when we have evidence for Nordic influence in Hungary, but have not been closely studied.

[2] *M.A.G.W.*, xxv, p. 40, fig. 56. From the Middle Bronze Age we have bits from the Italian *terremare*, from the settlement near Reichenhall in Bavaria (*Auh V.*, v, p. 396, fig. 2 *h*), from Denmark (*Aarbøger*, 1900, pp. 235 ff.), from Sweden and from Silesia (*P.Z.*, ii, pp. 173 ff., fig. 23).

of the European steppes. Did bands of Nordics, venturing into the passes of the Caucasus, glimpse from afar that Garden of Eden and tempted by its wealth make raids to the south? Did Sumerian merchants and explorers in search of the metal, timber, stone and gems, that their own alluvial land denied them, penetrate into the fastnesses of Armenia and beyond? Did Semites, descending the Halys from their colony in Cappadocia, take ship and cross the Black Sea? All such types of contact between Europe and Mesopotamia probably were in fact established. Rather later other visitants, coming this time by sea from the south-west, brought fresh ideas to the coasts of South Russia. Argonauts from the Cyclades, anticipating the exploits of the Milesians, undoubtedly founded trading colonies near the mouth of the Don as the tombs already mentioned show. Other mariners, "Children of the Sun," who found in the gems of the Caucasus the objects of their world-wide quest might have introduced the nomads of the steppes to the idea of the megalithic tomb and of divine kingship.

So our hypothetical Nordics in South Russia would have been less isolated than their distant kinsmen on the Baltic; among them the genesis of a vigorous neolithic culture would be easily comprehensible. They could have learned the simple neolithic arts of food-production and pottery-making; they could barter furs or the products of their herds for metal weapons and tools; in default thereof they could imitate such in stone and flint. And it is admitted on all hands that the Nordics in South Russia did absorb such influences. The view here expounded differs from that discussed in the last chapter only in this respect: whereas the Germanists recognize the elements derived from vase-painters, Mesopotamians and Aegeans only as secondary accretions on a Nordic culture brought fully-fledged from Scandinavia, it requires that the culture of the ochre-graves and the Nordic culture of Scandinavia itself should be fully constituted by the factors just enumerated. Our present hypothesis also pre-supposes migrations from the steppe not only to the south-east (Mesopotamia and Iran) and the south-west (Troy and the Balkans) but also to the north and north-west; the Germanists only admit the former movements.

Let us consider these points more closely. Note first that the ochre-graves are numbered by thousands; they presumably cover a considerable space of time. As a matter of fact Russian archaeologists have distinguished three phases of evolution in the Don-Donetz region.[1] The oldest graves are simple pits or stone cists containing very little metal and pots ornamented with linear designs produced by the impression of a cord. Next come chamber-tombs, called by the Russians "catacomb-graves", containing more metal and vases on which the cord-impressions form spirals and loops. Last of all come wooden coffins which overlap with the Iron Age.

The first phase would have on our view to correspond to the period of Aryan unity. By the second phase differentiation, marked by the growth of local styles in the pottery, had set in. Some families of pastoralists were deserting the steppe to adopt a sedentary life as cultivators in the fertile valleys that intersect it. It is in this phase that the Aegean influence is visible in the form of the tombs, and that of the vase-painters in the spiral decoration of the pottery. We might almost suspect an amalgamation between the peasants and the pastoralists, and from this time the valleys remained continuously occupied till the advent of the Scyths. Most Germanists will agree with this interpretation of the "catacomb-grave" period.

It will also be generally agreed that the Mesopotamian influence was most intense on the northern slopes of the Caucasus in the valleys of the Kuban and the Terek. Here truly royal graves contrasting in size and wealth with the poor *kurgans* of the steppe were reared. They are the tombs of chieftains who had led their followers on plundering expeditions into Armenia, Cappadocia and even Mesopotamia. The masses of gold and silver buried in these enormous barrows must partly be loot from the rich states south of the range. That is for instance manifest in the gold and silver lions and bulls that decorated the canopy under which one prince was laid to rest in the famous barrow near Maikop. These southern artifacts on the northern slopes of the Caucasus are the counterpart of the Caucasian objects which we met in North Syria in Chapter II. The raids that brought them north were the prelude to invasions. We may suspect that the ancestors of the Indians and the Iranians discovered as freebooters the roads that eventually led them to the throne of Mitanni and to the Indus valley.

[1] *Dawn*, pp. 143 f., Tallgren, *Götze-Festschrift*, p. 70, and Hubert Schmidt, *Vorgeschichte Europas*, p. 99, have recently expressed doubts as to the value of this chronological distinction.

Their advance cannot yet be followed in detail. We might suspect that the leader of the advance-guard of this invasion had wielded the copper battle-axe (axe-adze) found in a (?) barrow near Astrabad (p. 189). But one distinct, if undatable, migration round the Caucasus can be detected. It started north of the range, passed round the eastern flank of the chain and reached the Persian uplands west of the Caspian. The land-marks on its route are dolmenic tombs near Kala Kent[1] on the Baku peninsula and other sepultures explored by de Morgan in Talysh and Lenkoran.[2] The former group contained large spiral earrings with flattened ends of copper and beakers with analogies in the ochre-grave on the Kuban and the Dniepr. The tombs on the Caspian coasts of Transcaucasia are unmistakably allied in form to those of the Kuban valley,[3] but the furniture is less unitary. Copper pins with double or quadruple heads and copper battle-axes seem derived from more northerly types, but other ornaments and weapons must be referred to some still undated Mesopotamian culture. Finally iron objects have been found in some of the tombs in question, but are, according to de Morgan, due to a later wave of intruders. Hence the evidence for a movement of peoples from South Russia towards Mesopotamia is on the whole satisfactory. It may be significant that a fine stone battle-axe—the earliest dated specimen from this region—was laid in the foundation deposit of Shushinak's temple at Susa (VIIth century B.C.).

While some nomads were settling down in the valleys and others were constituting principalities on the slopes of the Caucasus, the remainder left upon the steppe would be forced to find outlets for their increasing numbers and fresh pastures for their growing herds by means of migration, just as the Danubian peasants had spread in Central Europe. But pastoralists do not spread slowly and regularly like cultivators but move rapidly by darts. Actual migration is preceded by exploratory expeditions in the summer, and such excursions reveal to the nomad other goals than mere grazing grounds—centres of wealth to be plundered and held to ransom. The enforced expansion from the steppe seems in fact to have been guided by some such ends.

[1] *Otchet*, 1897, pp. 141 ff.; Tallgren, in *F.M.*, 1924, p. 23, fig. 10.

[2] J. de Morgan, *Mission scientifique en Perse*, vol. iv; ibid., in *Mem. Dél. Perse*, viii.

[3] The double cist with a holed-stone for the transverse slab from Djönü (*Miss.*, loc. cit., fig. 48) is obviously a decadent variant of the dolmen of Tsarevskaya on the Kuban (*Dawn*, fig. 63).

One such wave of expansion will be admitted even by the Germanists.[1] It led the battle-axe folk to Troy and the east Balkans. The Trojan battle-axes (page 133) find their nearest parallels in South Russia. The route of this Nordic band would have lain along the open steppe north of the Pontus. A landmark on its route might be recognized in the treasure of Borodino in Bessarabia [2] which contained ceremonial axes of noble stone closely allied to those from Troy. The axes of the Trojan treasure must in any case be attributed to a chief who had come from the north coast of the Euxine. So too the stone battle-axes and flint celts which we noticed as intrusive elements in the settlements of the vase-painters in Bulgaria may be ascribed to a branch of the same stream of invasion and would mark the Aryanization of this end of the Balkan range.

We now turn to the westward movements. From the standpoint of this chapter the Nordics advancing on the Danube valley must have crossed the black-earth belt inhabited by vase-painters till about 1600 B.C. or later. The reader will, however, recall that the culture with painted pottery fell into two distinct periods in the Ukraine and Roumania (page 106). The older villages perished in flames and were not in all cases reoccupied, while Erösd, the cultural capital of the whole region, was finally ruined. This trail of fire and destruction might mark a first onslaught by nomads from the steppes; their goal would have been the Transylvanian gold fields. Monuments of their progress might be recognized in the copper axe-adzes that have been found in or near sites of the earlier peasant villages and above the ruins of Erösd. The later Hungarian axe-adzes that have such a wide distribution in Central Europe would then be the work of native metallurgists using local ores and working to the order of the new Nordic overlords.

Such an attack from the east would further explain very conveniently the phenomena we met in discussing the second phase of civilization in the Danube valley (page 151). If we assumed that some of the invaders from the steppes pressed on across the Alt into Hungary, sweeping along with them some of the conquered peoples of Transylvania, we could understand the Nordic skulls, the horse's bones, the copper axe-adzes and the barbarized painted

[1] So Tallgren, *Götze-Festschrift*, p. 75.

[2] *Materials for the Archæology of Russia*, xxxiv, pp. 1–14, pl. i. But this treasure, despite the close similarity of the battle-axes in it to those from Troy II, must really be later. It included a socketed spear-head that can hardly be older than the XVIth century B.C.

pottery that we met in the cemeteries of Lengyel, O Besseny and Lucskai. This wave of invasion would have been followed by others. One would have introduced the corded-ware, which we saw in the last chapter reached Hungary from the east, and the centre of which on our present view might be located between the Dniepr and the Don. And the various types of copper battle-axes which are concentrated in eastern Hungary and the present Roumania, but which extend to Bosnia, Dalmatia and Croatia, have been ascribed by Dr. Nagy to a series of invaders from the steppes.[1]

The connections between the Fatyanovo culture in Central Russia and the Copper Age ochre-graves further south are quite unmistakable. On the present thesis the former must be attributed to a movement of people up the Volga. It might even be argued that the same movement continued westward to Finland and Scandinavia.

Finally we come to the relations between South Russia and Scandinavia. The evidence for such relations is indisputable. On the view under discussion they must be explained by a multiplicity of waves and currents of migration ultimately converging upon centres of wealth—the amber deposits of Jutland, the Saale salt, the Elbe-Danube trade-routes. To unravel the complicated details of these movements here is frankly impossible. The pioneers would perhaps have been armed with polygonal battle-axes like Fig. 27, 5. After a pause in Little Poland some would have descended the Vistula and reached Jutland about the same time as the dolmen-builders. Others, going up stream, would have reached Silesia and then advanced as conquerors to the slopes of the Alps where they would have established themselves in the Copper Age hill-stations. Another band of invaders would have used a special type of globular amphora; such vases are common in Central Germany, Pomerania, Poland, Eastern Galicia and Poltava Government where they are regularly found in cist-graves accompanied by amber beads, but they are certainly connected with vessels found in an ochre-grave at Tsarevskaya on the Kuban. We should have to assume that this band was captained by a chief who had a tomb built for himself at Baalberg on the Saale in imitation of his ancestors' sepulchres at Tsarevskaya (p. 176).

But the most compact and ruthless body of invaders would

[1] *Dawn*, p. 188.

have been those who used cord-ornamented pottery. Their starting point would be near the Donetz valley where such pottery is found in the oldest class of barrows (p. 194) and whence their kinsmen would have set out for Transylvania. The northern bands would have aimed at Jutland and Thuringia. There they would emerge as the separate-grave folk and the Thuringian barrow-builders, whose subsequent wanderings westward were traced on p. 176.

Thus Kossinna's migrations would be reversed.

But is this reversal really feasible on the archaeological evidence? There are certainly arguments in its favour. It is a continuation of a drift which had begun in pre-neolithic times (p. 187). The typology of the battle-axes gives at least a satisfactory explanation for objects which are frankly puzzling in Scandinavia. It is supported by the fact that stone battle-axes of purely South Russian type have actually been found on the shores of the Baltic in Finland, Esthonia and even Denmark itself and that the concave chisels associated with the battle-axe cultures of Sweden seem to be derived from South Russian and ultimately Mesopotamian prototypes.[1] Still it would be unfair to allow the reader to infer that the vast mass of evidence patiently collected by all the leading authorities of Germany, Sweden, Poland and the Baltic States can so easily be dismissed.

A change in the direction of racial drift between pre-neolithic and late neolithic times is explicable in view of the deterioration in the climate of Scandinavia. At least by the full Bronze Age a current from Central Europe [2] was affecting South Russia and that continued till Scythian times. Most typological studies conducted by local archaeologists on the forms of tombs, celts, battle-axes and pottery and their distributions invariably give the priority to the Scandinavian and Central German forms.[3] The association of amber with globular amphorae in Polish and Galician graves does look as if their makers had come from the Baltic. And poor bone or clay pendants from tombs on the Kuban exactly resemble in shape amber ornaments from East Prussia.[4] Conversely the coloration of the skeleton with red

[1] Arne Europæus in *F.M.*, 1924, pp. 54 f.

[2] Tallgren in *Götze-Festschrift*, p. 76, n. 1, and in *S.M.Y.A.*, xxv, p. 94; cf. p. 203 below.

[3] This point is admirably illustrated by the maps in Kozłowski, *Młodsa*.

[4] Tallgren in *Götze-Festschrift*, p. 73.

ochre so characteristic of South Russia has only once been observed in the north—at Charlottenhohe in Uckermark.[1] The cumulative effect of the arguments here merely sketched is immense but not absolutely conclusive. The deciding factor must be chronology. Are any of the ochre-graves in South Russia really older than the earliest separate graves of Jutland (say 2500 B.C.)? [2] Are the double dolmens with globular amphorae at Tsarevskaya on the Kuban really older than their counterpart at Baalberg on the Saale (about 2000 B.C.)?

These questions can only be finally answered when the poor remains from the South Russian ochre-graves have been fully published and thoroughly studied. Professor Rostovtseff[3] on stylistic grounds dated the Copper Age tombs on the Kuban to rather before 2500 B.C. Professor Farmakovsky[4] on similar grounds arrived at a date quite a thousand years later. In the last few days the author has received a convincing study of the jewelry and implements from ochre-graves by Professor A. M. Tallgren. His conclusions are that the ochre-graves as a whole belong to the second millennium B.C., rather than the third. If this be correct, if these arguments are applicable not only to the "catacombs", but also to the earliest ochre-graves, then the attempt to reverse Tallgren's and Kossinna's migrations must be abandoned. The Nordic cultures in Jutland and Central Germany will be older than those in South Russia. The latter will not then be the monuments of the undivided Aryans, but only of a branch of that stock. The Aryanization of the Danube valley, the Alps and the Rhineland will be due to an expansion from the north, not an invasion from the east. The Nordic stone battle-axes will not be imitations of copper axe-adzes but must be derived from the horn implements with a hole for the shaft already in use at Maglemose, while the Hungarian axe-adzes may be due to trade with Crete. The battle-axe cultures of Jutland and Thuringia must have been generated out of some old native element through contact with the foreign civilization of the dolmen-builders. Their cord-ornamented vases must be the

[1] Schumann, *Die Steinzeitgräber der Uckermark*, p. 11.

[2] These dates are based on Sir Arthur Evans' chronology for Crete and consequently on Meyer's "short chronology" for Egypt. Should the new evidence in favour of the long chronology foreshadowed by Sir Flinders Petrie prove convincing, they must be substantially increased.

[3] Op. cit., pp. 20 ff.

[4] *Materials for the Archæology of Russia*, xxxiv, pp. 51 ff.; so Tallgren, *Götze-Festschrift*, p. 76; *S.O.F.*, i, pp. 339 f. See appendix to this chapter.

continuation of an older fabric, the roots of which Sophus Müller would trace back to the pre-dolmen age in Denmark.[1]

The present writer still thinks that the South Russian hypothesis, outlined in the preceding pages, may prove to be tenable; his confidence in it has, however, been shaken since he espoused it—with reservations—in an earlier work [2] by the appearance of the new articles by Kozłowski and Tallgren. In default of this only the Germanist theory is left.' The ochre-grave folk will still be Aryans but not the Aryans.

Conclusion

Aryan Groups in the Bronze Age

One question raised by our investigation must be left open to be settled by further researches on the South Russian material. But the vital point has emerged with perfect distinctness. The victorious expansion of the Nordic culture, whatever its origin, is the dominant fact of European prehistory from 2500 to 1000 B.C. The path of the prehistorian who wishes to draw ethnographical conclusions from archaeological data is often beset with pitfalls. The correlation of cultural with racial groups is generally hazardous and speculative. The diffusion of types and customs is as often due to trade and cultural borrowing as to movements of population; the infiltration of a new ethnic element need leave no mark on the external aspect of a culture. No such reservations impede the interpretation of the almost miraculous advance of the Nordic cultures. In their triumphant progress they repeatedly annexed regions previously occupied by higher types of culture. And such supersession of higher by lower is only explicable in racial terms.

Whether the Nordic culture originated on the shores of the Black Sea or of the Baltic its authors grew from an originally poor and insignificant group to the dominant power in the western world. By their pottery [3] and their battle-axes we can trace

[1] *Oldtidens Kunst i Danemarke*, i, p. 11. Cf. Menghin in Hoernes, pp. 736–40 and the illustrations there given.

[2] *The Dawn of European Civilization*, pp. 150, 206, 239, and 303.

[3] The pottery is admirably treated by Menghin, op. cit., pp. 734 ff.

them to the Rhine, to Switzerland, to Upper Austria, to Italy and to Hungary. They occupied the whole of the South Russian steppe and at least the one corner of Asia Minor that has been thoroughly explored. We can see them starting off across the Caucasus on the way to Mesopotamia and Iran. And the reasons for calling the Nordics Aryan are conclusive; wherever we can follow their movements in detail these Nordics formed the nuclei of cultural groups traceable in history as Aryan.

In southern Scandinavia and North Germany the battle-axe folk of the Stone Age had by the Bronze Age welded the composite population into a cultural unity; from this date the evolution of civilization in the North is self-contained and continuous.[1] It was therefore the work of the Teutons who inhabited those regions in the earliest historical times.

A kindred battle-axe folk from Thuringia had overrun the Rhineland, Wurtemburg and Switzerland during the latest Stone Age and Early Bronze Age there to conquer and mingle with Alpines, Prospectors and Danubians. Soon after the invaders reached the Rhineland, some of them, mixed with Prospectors, set off to invade Britain. But pottery and burial rites reveal that the Bronze Age barrows of the highland zone were built on the one hand by other descendants of the conquerors,[2] on the other by ancestors of a prominent element in the Iron Age population.[3] This composite population, dominated by battle-axe folk from Thuringia, must have been Celtic, since the western Hallstatt culture is as demonstrably Celtic as that of La Tène which originated within the same area.

Further south in Upper Bavaria and Upper Austria the fortified hill settlements like Altheim and the Copper Age lake-dwellings on the Attersee and Mondsee seem to have been founded by a Nordic aristocracy ruling over an aboriginal Alpine stock; the battle-axes, like Fig. 27, 5, and the pottery[4] suggest that these rulers had come immediately from Silesia across Moravia. A southward movement on the part of this mixed stock will perhaps best account for some elements in the *terremare* of Italy. The

[1] *Dawn*, pp. 214 ff.

[2] The late barrows with cord-ornamented ware in Wurtemburg are parallel to the Early Bronze Age cemeteries in the valleys; their continuity with the Middle Bronze Age barrows is shown by the pottery; vases such as Behrens, pls. x, 9, xvii, 3 and 12, xviii, 1, are plainly derived from cord-ornamented ware.

[3] For the survival of Bronze Age ceramic styles into the Iron Age see Behrens, p. 218, and Schumacher in *Auh V*, v, pl. 40 and text.

[4] On this point see Menghin, op. cit., pp. 762–5.

intimate connection between Celtic and Italic would then be explained not only by the contiguity of their centres of dispersion but also by the Alpine element common to the speakers of both groups of languages.

Between the Elbe and the Oder the Aunjetitz culture of the Early Bronze Age resulted from a fusion of Nordics, Prospectors and Danubians; the Nordic element is betrayed both by the pottery and the stone battle-axes—of markedly East European type—found even in Bronze Age graves. From the Aunjetitz culture sprang a group of allied cultures represented by fields of cinerary urns of what is called the Lausitz type.[1] From Silesia and Poznania these urn-fields spread in the Late Bronze and Early Iron Ages into Bohemia on the west and far across Poland and Galicia on the east. Controversy still rages as to who were the authors of this culture, but the most probable view is that it was created by the Slavs.[2]

Whether the ochre-grave folk were native to South Russia or immigrants from the North, we can see more than one Aryan nation growing out from them. As already indicated, the royal graves of the Copper Age on the Kuban and the Terek were built by ancestors of the Indo-Iranians. From the opposite end of the Pontic steppe issued those who were to become Hellenes in the Balkans. Thus, even if the ochre-graves were not built by the original Aryans, the ancestors of the Greeks and the Indo-Iranians would alike have sojourned for a while in South Russia, and been exposed to the same foreign influences there. That circumstance might help to account for the similarity between the Copper Age of South Russia and the assumed proto-Aryan civilization; for our picture of the latter is, in many of its details, based on equations common only to Greek and Indo-Iranian.

It is perhaps premature to designate Phrygian the Trojan prince, whose dominion was symbolized by splendid battle-axes of noble stone. That he and his followers came from the Pontic steppe is in any case certain and the results of Chapter V require that he should have been Aryan. Again the result of the amalgamation between the vase-painters of Bulgaria in the Copper Age and

[1] Recently demonstrated by Baron von Richthofen, *Mannus*, 1925 (Erganzungsband), pp. 140 ff.

[2] So Pic, *Die Urnengräber Böhmens*, and, more recently, Kostrzewski, *Wielkopolska w czasach przedhistorycznych*, Posen, 1923. Kossinna assigns the Lausitz culture to the Thracian Carpodoki (*Mannus*, xi–xii, pp. 232 ff., etc.); Schuchhardt's ascription of it to Germans (*Alteuropa*) is universally rejected even in Germany.

intruders from the steppes was to make the former Thracians. At the same time other vase-painters in South Russia had been absorbed by the local Nordic population, for the pottery from the later ochre-graves (catacomb-graves) in its spiral decoration and other features betrays influence from the black-earth zone. The sedentary population resulting from this fusion seems to have remained on the coasts of the Black Sea till the advent of the Scyths. It may then be called Cimmerian. At the same time it included the same constituents as the people of Thrace proper, so it may in a sense be designated Thracian.

Here I must again insist that the habitation of the ochre-grave area was continuous till Scythian times. The Cimmerians probably formed the sedentary element dwelling on the coast and in the valleys; in the hinterland there remained more nomadic Nordic tribes. Their domain would by now have embraced Transylvania, since the vase-painters ultimately vanish from the Ukraine and Nordic barrows overlie the ruins of their villages while kindred barrows are met in Transylvania. These nomad hordes were thus in contact with the progressive bronze industry of the Danube valley and so were well fitted to act as vehicles in the eastward diffusion of Central European metal-work. That diffusion was beginning in the Early Bronze Age (about 1800–1500 B.C.) when penanular bracelets with recoiled ends of Hungarian-Bohemian type appear in catacomb graves of Tauric Government.[1] It was continued in the Late Bronze Age when a regular secondary focus of the Hungarian Bronze Age culture arose in the Ukraine and Hungarian types of socketed celts and kindred implements were spread as far as Central Russia, the Urals and Siberia.[2] The persistence on the steppe of a mobile population in touch with the centres of Danubian metallurgy enables us to understand the western relations of the Koban culture discussed on pp. 117 f. above. One such tribe, migrating to the Caucasus in the wake of the Indo-Iranians, perhaps under the pressure of the advancing Lausitz people, might have brought with them the purely European objects of Late Bronze Age type found in the Koban cemetery. Incidentally it may be noted that the Koban skulls are mesaticephalic and very different from the extremely broad-headed skulls of the native Transcaucasian tribes.[3] It is possible

[1] Tallgren in *Götze-Festschrift*, p. 76, n. 1; for the type see *Dawn*, fig. 91, 11–12.
[2] Tallgren, l.c., p. 74; cf. *S.M.Y.A.*, xxv, p. 94; *S.O.F.*, i, p. 339.
[3] de Morgan, *Caucase*, pp. 203 ff.

that this movement brought the Aryan nucleus of the Armenians to the northern slopes of the Caucasus.

Finally we may observe that certain copper objects of typically Russian form have been found sporadically as far east as Eastern Turkestan just as socketed celts with Scandinavian and Hungarian analogies are common throughout Siberia.[1] It is frankly difficult to say to what extent these objects were diffused by the rude hunting tribes of the northern forest which we connect rather with the ancestors of the Finns than with the Aryans. It is nevertheless possible that the Tocharians were among the tribes that drifted eastward across the mysterious steppes and deserts of Central Asia.

Thus the great majority of the Aryan nations of historical times can be shown to be descended from the Nordic battle-axe folk of the Stone Age. By the aid of pottery and weapons [2] they can be traced back with more or less certainty to one of two centres—South Russia or Scandinavia. The first business of future researches must be to determine which of these really has the priority. A complementary task is to unravel the cultural tangle still presented by Hungary, the north-west Balkans, and Iran. The precise links to connect the most important of all Aryan nations—the Greeks, the Iranians and the Indians—with one another and with their brothers have at present to be inferred; they must be found in these regions.

APPENDIX TO CHAPTER VIII

In view of the critical importance of the date of the ochre-grave culture in South Russia, I add a list of objects from such graves to which more or less accurately datable parallels are known elsewhere.

Maikop (Kuban)—razor (?) (*Dawn*, fig. 61, top left): cf. Mochlos tomb iv, MM. III (1700–1600 B.C.). (Seager, *Mochlos*, fig. 45.)

[1] *S.M.Y.A.*, xxv, p. 123, fig. 71; Minns, *Scythians and Greeks*, pp. 241–6.

[2] To infer a migration from the distribution of weapon types alone would, I submit, be rash. The diffusion of such objects is very often due to trade. Only in a few exceptional cases, particularly in early periods or when the articles are of a more archaic character than their context, such as the Trojan battle-axes, can such commercial diffusion be ignored. It is quite a different matter when we have a whole cultural complex moving about from place to place as happens with the battle-axe cultures. It is for this reason that I have been unable to follow Mr. Peake, who traces the migrations of the Western Aryans from the Danube valley by means of swords.

Tsarevskaya (Kuban)—poker-butted spear-head (*Dawn*, fig. 62; cf. Early Hittite graves near Carchemish (? 1900–1750 B.C.). (*L.A.A.A.*, vi, pl. xix, c. 4.)

Tsarevskaya (Kuban)—? dagger with bronze hilt (ib.); cf. Italian Aunjetitz daggers of Central Europe (1750–1450 B.C.).

Konstantinovka near Novocherkask (Don)—winged beads (*Dawn*, fig. 65, 2); cf. phallic beads, Paros, ? E.M. III (2400–2100 B.C.). (Ib., fig. 20, 3.)

Or cf. Egyptian "fly" ornament (*Menat*), early XVIIIth Dynasty (XVIth century B.C.).

Same grave—"papyrus staff" amulet (also from grave II*b* near Konstantinovka, Terek, Tallgren, *Götze-Festschrift*, fig. 1); cf. similar beads from Paros, same date (l.c., fig. 20, 4), or Egyptian amulets of various dates.

Novogrigoryevka (Dniepr, catacomb-grave) ? segmented bead of bone (Tallgren, l.c., fig. 13); cf. segmented stone beads Vrokastro E.M. II (2800–2400 B.C.) or ditto, paste Assur, before 2500 B.C. (?) (Andrae, fig. 61) or ditto paste, Crete, MM. III, and later.

Same grave—copper disc with punctured ornament (l.c., fig. 12); cf. disc from Stollhof in Lower Austria, c. 1800 B.C.), or

Same grave and often in other tombs—hammer-headed pins (*Dawn*, fig. 65, 4–6); cf. Remedello, silver (? 2000 B.C.) or Kazbek, bronze (Tallgren, *S.O.F.*, i; p. 327). (1200–1000 B.C. ?) or Argive Heræum, Geometric (900–800 B.C.), (Waldstein, *The Argive Heræum*, pl. lxxx, 353–364). The last parallel is very close.

Jackowice near Kiev—helical copper earrings with flattened ends (*Swiatowit*, vi, fig. 26); cf. earrings from treasures at Troy, II, i, and from Central European Aunjetitz graves.

Tallgren further compares the hollow hemispherical "beads" of copper from Remontoye (Astrakhan) and Kru (Kuban) with rather similar hollow beads from Hungary (end of Early Bronze Age, say XVIth century). But these objects are in reality only the metal covers for buttons of some perishable substance. Beads of the type they presuppose are met at Anau in Culture I (Pumpelly, op. cit., fig. 295), and similar covers in gold in E.M. III or MM. I tholoi in Crete (Xanthudides, *Vaulted Tombs of the Mesara*, pl. lxii).

Arguments based on typological studies of battle-axes have been omitted as deceptive—note that fine stone battle-axes are actually met in Scythian graves in South Russian (*Otchet*, 1899, pp. 47-8) and in Iron Age settlements and barrows in Bosnia (*W.M.B.H.*, i, p. 40, fig. 23; iv, p. 6, fig. 11)—as have comparisons with long-lived Asiatic types, such as axes, spear-heads, and forks which are chronologically worthless. Two further points must be stressed. Although the ochre-graves are very numerous, the internal development illustrated by their furniture is so slight that it is difficult to spread them over eighteen centuries. Secondly Scythian graves quite often occur in the same group of barrows as ochre-graves, in fact in some cases the two types of interment are met in the same barrow, though the Scythian remains seem invariably to lie at a higher level.

Such considerations are far from decisive. It must, nevertheless, be conceded that their cumulative effect favours a relatively late date for the ochre-grave culture. It fits best into the general chronological framework of European pre-history as viewed from the standpoint of Central Europe if placed in the second millennium B.C. At the same time it may be that the rude products from the ochre-graves to which Bronze Age parallels from Central Europe have been cited should be looked upon as the prototypes of forms which, under the stimulus of the amber trade, were fruitfully elaborated there.

Socketed spear-heads have been found in ochre-graves in the Terek excavated in 1925.

CHAPTER IX

THE RÔLE OF THE ARYANS IN HISTORY

The reader may think that the rôle here assigned to the Aryans is an extremely modest one. If the view advanced in the last chapter be correct, they were not the inaugurators of the neolithic civilization even in Europe nor were they as a whole the pioneers in the use of bronze or iron. The makers of the kitchen-middens on the Danish coasts have been justly termed " disgusting savages ". Even stronger epithets might be applied to the other claimants to the title of proto-Aryans; for a suspicion of cannibalism clings to the ochre-grave peoples.[1] Even in barbarian Europe the material culture of the Nordics was not originally superior to that of the Danubian peasants or the megalith-builders; in Transylvania they appear frankly as wreckers; in the Ancient East and the Aegean they appropriated and for a time impaired older and higher civilizations.

It was perhaps something to be able to rise from a state of wretched savagery even to overcome more civilized tribes. Not all savages know how to take advantage of the gifts of traders as the Nordics did on the Baltic and in South Russia. But what was their positive contribution to the capital of human progress? We may at least say that they were not merely destroyers. They knew how to profit by and improve on the achievements of their victims. From the fields they had wasted choicer blossoms grew.

To appreciate that we should have to proceed by way of contrast. Only a few points can be suggested here. In Hither Asia civilization had reached an exceptionally high level by the IVth millennium before our era. As the earlier achievements of the Sumerians gradually become better known and provide a standard of comparison, we begin to feel how relatively trifling were the advances made during the next two thousand years. The metal work of the First Dynasty of Ur reveals a perfect mastery over elaborate technical procedures. The chief types of tool unearthed in the ruins of Assyrian cities had been already in use under the early kings of Kish. From an aesthetic standpoint the copper heifers

[1] Tallgren, in *Götze-Festschrift*, p. 69, n. 1.

and shell inlays from A-anni-padda's temple at Tell el 'Obeid (about 3200 B.C.) are already masterpieces breathing a delicate feeling for life unsurpassed till the Persian period. Even the marvellous lions and horses on the Assyrian bas-reliefs represent a comparatively insignificant advance. Such improvements in arts and crafts as are detectable in Assyrian civilization are generally attributed to the influence of the Hittites, among whom some sort of Aryan inspiration was certainly at work. In the political domain progress had been more substantial at least in the IIIrd millennium. The Semite Narâm-Sin had made a step in the direction of unity, and Hammurabi carried his work a stage nearer completion. But Hammurabi's empire only imposed peace and political solidarity on the disunited Orient for a brief period and his wise laws, themselves largely derived from much older statutes, were in the sequel altered for the worse. The Assyrians, who were the ultimate heirs of Babylonian sovereignty, added nothing to the political capital of mankind. Their empires were indeed vaster than anything that had preceded them, but they rested upon naked force and unmitigated cruelty and failed to confer on the subject peoples durable peace or lasting security in return for crushing tributes preceded by pitiless devastations. The Assyrians forged a terrible engine for plunder and extortion; for a governmental structure to shelter trade and intellectual intercourse we have to await the Aryan Darius. Finally the religious ideas current throughout the Ancient East remained utterly primitive and showed not the least development in the direction of moralization or genuine monotheism till the VIth century. The sole exception was in the reign of Amenhotep IV in Egypt, and it has been pointed out that the world's first heretic was brought up at a Court where Mitannian princesses played a prominent part and his cult of the solar disc has been thought to reflect Aryan inspiration.

The accession to power of the Iranian Achaemenids brought in its train an aesthetic, political and religious revolution. No doubt the Persians had the benefit of the experience of their predecessors. The transformation achieved is none the less startling. Achaemenid art is characterized by a sobriety and verisimilitude unknown in Hither Asia since the Sumerian period, and yet incorporated all the technical improvements of the intervening centuries. Most striking is the lively individuality of the human figures as contrasted with the stiff and expressionless types of the Hittites, Babylonians or Assyrians. This seems an essentially Aryan trait. It is further very significant that the Persians should at once have proceeded to create a simple and almost alphabetic syllabary out of the clumsy and incredibly complicated cuneiform script which Sumerians, Hittites, Assyrians and Chaldaeans in turn had been content to use with no sensible modification for two thousand years. The Persian Empire was not only infinitely wider than even the greatest domains conquered by a Sargon or a Sennacherib. It was organized with statesmanlike genius by the great Darius and for two hundred years brought peace to the war-scarred lands of the Near East. Under its shelter merchants and philosophers could travel unhampered from the Aegean to the Indus; its royal roads were arteries along which not only military force but also the scientific and religious inspiration of the Ancient East flowed to Greece and Rome. That enlightened and prudent organization, contrasting so strongly with the plundered wastes dominated by Assyrian despots, was planned by Aryan princes and administered by Aryan governors. Its official religion, Zoroastrianism, was inspired by genuinely moral conceptions and was international in its appeal and monotheistic in its essence even though that internationalism and monotheism may in a sense reflect the imperialism of its royal votaries.

In Hellas the work of the Aryan invaders is less easily recognizable. The Minoans had created a civilization which was truly European and an art which, at its best, surpassed any contemporary product of the East. But that civilization seems to have lacked the vigour for expansion; it was already in its decline when the Achaeans overthrew it. To Egypt the Minoans brought tribute or gifts; the Achaeans slashing swords. Minoan merchantmen reached the Delta and the Levant; men of war were captained by the Hellenes. Minoan art reached its zenith by 1600 B.C.; in the Second Late Minoan Age conventionalization had set in to lead to decadence in the Third. That decadence was not arrested by the infiltration of Achaean dynasts, but they at least inspired new decorative principles which bore abundant fruit in the VIth century. The metopic style in ceramic art provided a corrective to the luxury and exuberance of Minoan decoration which still retained something Oriental. The Aryan interest in humanity provided the potter with a theme in which, after the rude attempts of the warrior-vase and the Dipylon, his classical successors were to achieve supremacy.

But it is in continental Europe that the work of the Aryans as founders of Western Civilization is most readily apparent. The west and north of our continent had been occupied by the megalithic peoples. The stupendous size of these monuments and the skill employed in their erection betoken a relatively high civilization in their builders. If Mr. Perry be right, the founders of this civilization were equipped with all the material and intellectual resources of the Egyptians of the pyramid age. Yet in France and the Iberian peninsula this civilization shows not a trace of internal development, not a vestige of progress. Though the number of the monuments was multiplied indefinitely, their furniture remained rude and barbaric. Despite a favoured situation in metalliferous regions and the fertilizing influences born along the western trade routes, the megalith-builders continued to use flint and stone or at best copper when other peoples were working bronze and iron. It is scarcely possible to point to a single fruitful type of tool or ornament which originated in the megalithic regions of France or Portugal. It seems as if these people were wholly absorbed in the cult of the dead and as if superstitious observances monopolized and paralysed all their activities. Complete stagnation ruled in industry, and to find parallels to their culture we have only to visit the Pacific Islands which have been exposed to a similar influence. This civilization which stagnated on the Atlantic coasts for a thousand years or so, from the latter half of the IIIrd millennium B.C., was not European; Western civilization was brought to the West by the Celts from Central Europe towards the end of the IInd millennium!

The prehistory of Britain is very different. Soon after 2000 B.C. a battle-axe people conquered the territory previously occupied, as in France and Portugal, by the megalith builders. With the advent of the invaders a period of rapid and original development set in. The rich and varied furniture of the intruders' round barrows is in striking contrast to the monotonous poverty of the grave goods from the older long barrows. We know now that the battle-axe wielders were admixed with Aryans, and the truly Western civilization which henceforth ruled in Britain was obviously promoted by them.

In Scandinavia the contrast to France and the Iberian peninsula is even more fundamental. Here, too, men built megalithic graves, but their furniture here is totally different to anything discoverable further west. And besides the megalithic tombs were other graves covering the remains of a people, who, whether they were come from

South Russia or represented a section of the pre-dolmenic population, were, we believe, Aryan in character. It was these who inspired the higher developments even in the megalithic culture of the North. The interaction of the two types of civilization was the mainspring of a rapid progress. And ultimately the division was overcome; the Aryans imposed their authority and their culture—partly, if you will, a borrowed culture—on the whole region, welded the disparate racial groups and the scattered clans into a national unity in which western and eastern ideas were blended to an European whole and called forth a progressive society no less brilliant in trade and art than in war. The gulf between French and Scandinavian culture at the beginning of the IInd millennium is enormous. The superiority of the former is the measure of the contribution made by the Aryan element to European civilization.

In the Danube valley the tale is not very different. The early peasants had reached no mean level of culture. The material additions introduced by the Nordic infiltrations and conquests were of secondary importance. Often indeed these intrusions actually caused a set-back to material civilization. But the first culture was essentially a peasant civilization and as such unprogressive and rigid. Left to itself it might have remained on the level of a totemic society in Melanesia or North America. In out of the way corners Danubian I culture did actually persist for a long time in a fossilized condition. But just where the Nordic invasions had been most persistent we find a Bronze Age art and industry which are truly European in their originality. The ferment which transmuted the societies of agricultural clans into the heroic tribes of the Bronze and Iron Ages, thus opening the way to initiative and individuality, we regard as Aryan.

Thus the Aryans do appear everywhere as promoters of true progress and in Europe their expansion marks the moment when the prehistory of our continent begins to diverge from that of Africa or the Pacific.

Perhaps disappointment has now given place to bewilderment in the reader's mind. How precisely did the Aryans achieve all this? It was not through the superiority of their material culture. We have rejected the idea that a peculiar genius resided in the conformation of Nordic skulls. We do so with all the more confidence that, by the time Aryan genius found its true expression in Greece and Rome, the pure Nordic strain had been for the most part absorbed in the Mediterranean substratum: the lasting gift bequeathed by the Aryans to the conquered peoples was neither a higher material culture nor a superior physique, but that which we mentioned in the first chapter—a more excellent language and the mentality it generated. It is particularly significant that where, as in Mitanni, the Indo-European language was not retained, the effects of an infusion of Aryan blood did not come to fruition.

At the same time the fact that the first Aryans were Nordics was not without importance. The physical qualities of that stock did enable them by the bare fact of superior strength to conquer even more advanced peoples and so to impose their language on areas from which their bodily type has almost completely vanished. This is the truth underlying the panegyrics of the Germanists: the Nordics' superiority in physique fitted them to be the vehicles of a superior language.

BIBLIOGRAPHICAL NOTE

PERIODICALS

The following abbreviations have been used in citations:—

AfA.	*Archiv für Anthropologie* (Brunswick).
A.M.	*Mitteilungen der kaiserlich deutschen archäologischen Instituts, athenische Abteilung.*
Ant. J.	*Antiquaries Journal* (London).
Arch.	*Archæologia* (London).
'Αρχ. Δελτ.	*'Αρχαιολογικὸν Δελτίον* (Athens).
B.P.	*Bullettino di paletnologia italiana* (Parma).
B.S.A.	*Annual of the British School at Athens.*
B.S.J.	*Bulletin of the British School at Jerusalem.*
B.S.R.	*Papers of the British School at Rome.*
C.Q.	*Classical Quarterly* (London).
C.R.	*Classical Review* (London).
'Εφ. 'Αρχ.	*'Αρχαιολογικὴ 'Εφημερίς* (Athens).
F.M.	*Finskt Museum* (Helsingfors).
Glotta	(Gottingen).
I.F.	*Indogermanische Forschungen* (Strasburg).
I.J.	*Indogermanisches Jahrbuch* (Strasburg).
J.A.O.S.	*Journal of the American Oriental Society* (Boston).
J.E.A.	*Journal of Egyptian Archæology* (London).
J.H.S.	*Journal of Hellenic Studies* (London).
J.R.A.I.	*Journal of the Royal Anthropological Institute* (London).
J.R.A.S.	*Journal of the Royal Asiatic Society* (London).
K.Z.	Kuhn's *Zeitschrift für vergleichender Sprachwissenschaft* (Berlin).
L.A.A.A.	*Liverpool Annals of Anthropology and Archæology.*
L'Anthr.	*L'Anthropologie* (Paris).
M.A.G.W.	*Mitteilungen der anthropoligischen Gesellschaft in Wien.*
M.A.G.Z.	*Mitteilungen der antiquarischen Gesellschaft in Zurich.*
M.D.O.G.	*Mitteilungen der deutschen Orient-Gesellschaft* (Berlin).
Mannus	(Berlin).
Mem. Del. P.	*Mémoires de la délegation en Perse* (Paris).
O.L.Z.	*Orientalische Literaturzeitung* (Berlin).
P.S.B.A.	*Proceedings of the Society of Biblical Archæology* (London).
P.Z.	*Prähistorische Zeitschrift* (Berlin).
S.O.F.	*Studia Orientalia Fennica* (Helsingfors).
S.M.Y.A.	*Suomen Muinaismuistoyhdistyksen Aikakauskirja* (Helsingfors).
Syria	(Paris).
W.M.B.H.	*Wissenschaftlichen Mitteilungen aus Bosnien und Herzegowina* (Vienna).
W.P.Z.	*Wiener prähistorische Zeitschrift.*
W.V.D.O.G.	*Wissenschaftliche Veröffentlichungen der deutschen Orient-Gesellschaft* (Berlin and Leipzig).
Z.D.M.G.	*Zeitschrift der deutschen Morgenlandgesellschaft* (Berlin).
ZfE.	*Zeitschrift für Ethnologie* (Berlin).

BOOKS

Works frequently cited as authorities alone are here included. On the other hand, without attempting to give a complete bibliography of the literature of the Aryan controversy, some of the older works of especial interest for the history of the question are given below, even though they be quite out of date from a scientific point of view.

Åberg, Nils . . *Das nordische Kulturgebiet* (Uppsala, 1918)

Allen, T. W. . . *The Homeric Catalogue of Ships* (Oxford, 1921).
Homer : The Origins and Transmission (Oxford, 1924).

Bechtel, F. . . *Die griechische Dialekte* (Berlin, 1921 f.)

Bender, H. H. . *The Home of the Indo-Europeans* (Princetown, 1922).

Behrens, G. . . *Bronzezeit Süddeutschlands* (Mainz, 1916).

Cambridge Ancient History (cited *C.A.H.*).

Cambridge History of India (cited *C.H.I.*).

Chadwick, M. . . *The Heroic Age* (Cambridge, 1912).

Chantre, E. . . *Mission en Cappadoce* (Paris, 1883).
Recherches anthropologiques dans le Caucase (1885–7).

Childe, V. Gordon . *The Dawn of European Civilization* (London, 1925).

Dechelette, J. . . *Manuel d'archéologie préhistorique celtique et gallo-romaine* (Paris, 1908–1914).

Delaporte, L. . . *Mesopotamia : Babylonian and Assyrian Civilization* (London, 1925).

Feist, S. . . . *Kultur, Ausbreitung und Herkunft der Indogermanen* (Berlin, 1913).

Fick . . . . *Vorgriechische Ortsnamen* (Gottingen, 1905).

Forsdyke, E. J. . *Catalogue of Greek and Etruscan Vases in the British Museum*, vol. i, i (London, 1925).

Frazer, Sir James . *Lectures on the Early History of the Kingship* (Cambridge, 1910).

Giles, P. . . . See *C.A.H.* and *C.H.I.*

Götze-Festschrift . *Studien zur vorgeschichtlichen Archäologie Alfred Götze . . . dargebracht . . .* (Leipzig, 1925).

Haddon, A. C. . . *The Races of Man* (Cambridge, 1924).

Hall, H. R. . . *The Ancient History of the Near East* (London, 1913).
The Oldest Civilization of Greece (London, 1901).

Hirt . . . . *Die Indogermanen* (Strassburg, 1905–7).

Hoernes, R. . . *Urgeschichte der bildenden Kunst in Europa* (Vienna, 1924).

von Ihering, R. . . *The Evolution of the Aryans* (trans. London, 1897).

Keith, Sir Arthur . *The Antiquity of Man* (London, 1925).

Keller and Reinerth . *Urgeschichte des Thurgaus* (Frauenfeld, 1925).

Kossinna, G. . . *Die Indogermanen* (Würzburg, 1921). (See also note on p. 168.)

Kretchmer, P. . . *Einleitung in die Geschichte der griechischen Sprache* (Gottingen, 1896).

de Lapouge . . *L'Aryen* (Paris, 1899).

Latham . . . *Elements of Comparative Philology* (London, 1862).

Macalister . . *Excavations at Gezer.*

MacDonell, A. A., and Keith . . . *Vedic Index of Names and Subjects* (1907).

Menghin, O. . . See Hoernes.

Meyer, E. . . . *Reich und Kultur der Chetiter* (Berlin, 1914).
de Michelis, E. . . *L'Origine degli Indo-Europei* (Turin, 1903).
Minns, E. H. . . *Scythians and Greeks* (Cambridge, 1911).
Moret and Davy . *From Tribe to Empire* (London, 1925).
de Morgan, J. . . *Mission au Caucase* (Paris, 1889).
Mission scientifique en Perse (Paris, 1894).
Prehistoric Man (London, 1924).
de Mortillet . . *Formation de la nation française* (Paris, 1897).
Müller, F. Max . . *Bibliographies of Words and the Home of the Aryans* (London, 1888).
Pargiter . . . *Ancient Indian Historical Tradition* (Oxford, 1922).
Peake, Harold . . *The Bronze Age and the Celtic World* (London, 1922).
Peet, E. . . . *The Stone and Bronze Ages in Italy and Sicily* (Oxford, 1909).
Penka, Carl . . *Origines Ariacæ* (Vienna, 1883).
Die Herkunft der Arier (Vienna, 1887).
Penrose, Harland . *The Peloponnesos in the Bronze Age* (*Harvard Studies in Classical Philology*, 1923).
Perry, W. J. . . *The Growth of Civilization* (London, 1923).
Pittard, E. . . *The Races and History* (London, 1925).
Petrie, Sir William M. Flinders . . *A History of Egypt* (London).
Poesche . . . *Die Arier* (Jena, 1878).
Pumpelly, R. . . *Explorations in Turkestan* (Carnegie Publications, No. 73).
Randall-MacIver . *Villanovans and Early Etruscans* (Oxford, 1924).
Ridgeway, Sir William *The Early Age of Greece* (Cambridge, 1901).
Ramsay Studies . . *Anatolian Essays and Studies presented by Sir William Ramsay* (1924).
Rostovtseff . . *Iranians and Greeks in South Russia* (Oxford, 1922).
Sayce, A. H. . . *Introduction to the Science of Language* (London, 1880).
Schrader, O. . . *The Prehistoric Antiquities of the Aryan Peoples.* Translation by Jevons (London, 1898).
Reallexikon der indogermanischen Altertumskunde (1st edition, 1902; 2nd edition edited by Nehring in progress; letters *A* to *M* have so far appeared).
Sergi, G. . . . *Gli Arî in Europae in Asia* (Turin, 1902).
Taylor, Isaac . *The Origin of the Aryans* (London, 1889).
de Ujfalvy . . *Les Aryens au nord et au sud de l'Hindoukouch* (Paris, 1896).
Vendryes . . . *Language* (London, 1925).
Wilke, G. . . . *Südwesteuropäische Megalith Kultur und ihre Beziehungen zum Orient* (Würzburg, 1912).
Zaborowski . . Articles in *Revue de l'école anthropologique de Paris*, especially 1898.
Zimmer, H. . . *Altindisches Leben* (Berlin, 1879).

INDEX

Printed in Great Britain by Stephen Austin & Sons, Ltd., Hertford.